AQA GCSE Modular Maths

Intermediate: Module 5

Trevor Senior and Gordon Tennant

Contents

Pearson Education Limited
Edinburgh Gate
Harlow
Essex
CM20 2JE
England

www.longman.co.uk

First published 2003

ISBN   0 582 79595 8

Design and typesetting by Mathematical Composition Setters Ltd, Salisbury, Wiltshire

Printed in the U.K. by Scotprint, Haddington

The publisher's policy is to use paper manufactured from sustainable forests.

Acknowledgements

The publisher would like to thank Keith Gordon and Tony Fisher for their advice on the manuscript.

We are grateful for permission from the Assessment and Qualifications Alliance to reproduce past exam questions. All such questions have a reference in the margin. AQA can accept no responsibility whatsoever for accuracy of any solutions or answers to these questions. Any such solutions or answers may not necessarily constitute all possible solutions.

Introduction

AQA GCSE Modular Maths, Module 5 is written by experienced examiners to help you get the most out of the Number; Algebra; and Shape, Space and Measures element of your Modular Mathematics course. This is part of a series of course books that cover the AQA Specification B at Intermediate Tier.

This book can also be used to provide the background towards the *AO1 Using and Applying Mathematics* coursework task. It will also support the Number; Algebra; and Shape, Space and Measures topics for other GCSE Mathematics courses.

This book delivers the Module 5 specification in 40 chapters. The general areas of these include:

- Revision of number skills encountered in Module 3, extending to more demanding number skills
- Introduction to Algebra, extending to more advanced algebraic skills needed for solving various types of equation (linear, simultaneous, quadratic)
- Shape, Space and Measures including parallel lines, transformations, Pythagoras' theorem and trigonometry

Each chapter has a short introduction followed by:

- *Examples and Solutions*
- *Practice questions*
- *Practice exam questions.*

Some of the practice exam questions are past exam questions.*

Answers are provided for all of the *practice* and *practice exam questions.*

Examiner tips, located in the margin, give useful hints and advice such as which topics may appear in the calculator and/or non-calculator section of the exam.

Reminders, also located in the margin, give key knowledge already covered earlier in the book with references to other parts of the book.

At the end of the book there is a Practice exam paper. The Module 5 exam has two papers: Paper 1 (non-calculator) and Paper 2 (calculator). These will help you prepare for your Module 5 exams.

Good luck in your exams!

Trevor Senior and Gordon Tennant

* Past exam questions are followed by a reference in the margin containing the year and awarding body that set them. All exam questions are reproduced with kind permission from the Assessment and Qualifications Alliance. Where solutions or answers are given, the authors are responsible for these. They have not been provided or approved by the Assessment and Qualifications Alliance and may not necessarily constitute the only possible solutions.

1 Revision of number skills

In Module 3 you will have already met the ideas of:

- squares
- square roots
- cubes
- cube roots
- rounding:
  - decimal places
  - significant figures
  - appropriate degree of accuracy
- standard (index) form, expressed in conventional notation and on a calculator display
- percentage of a quantity.

Squares

The topics in this chapter can be examined in Module 3 and Module 5.

A **square number** is a number multiplied by itself. For example, the square of 5, written 5^2, is equal to $5 \times 5 = 25$.

In the exam, you will be expected to know the squares of all the integers from 1 to 15.

Negative numbers can also be squared. When you square a negative number, the result is always positive, e.g. $-3 \times -3 = +9$ or $(-3)^2 = 9$.

Number	Square of number
1	1
2	4
3	9
4	16
5	25
6	36
7	49
8	64
9	81
10	100
11	121
12	144
13	169
14	196
15	225

Square roots

The **square root** of a number, written $\sqrt{\ }$, is the value which when squared gives the number itself.

Finding the square root of a number is the reverse of squaring a number.

Numbers have both **positive square roots** and **negative square roots**. You can write the square root of 25 as $\sqrt{25}$. Since $5 \times 5 = 25$, the numerical value of $\sqrt{25}$ is 5. Since $-5 \times -5 = 25$, -5 is also a square root of 25.

Every square root has two numerical values of the same size; one is positive and one is negative.

In general, you can assume that you are using the positive square root in a question unless it asks for two answers.

Number	Square roots of number
1	1 or −1
4	2 or −2
9	3 or −3
16	4 or −4
25	5 or −5
36	6 or −6
49	7 or −7
64	8 or −8
81	9 or −9
100	10 or −10
121	11 or −11
144	12 or −12
169	13 or −13
196	14 or −14
225	15 or −15

Estimating values of square roots

In the exam, you may be asked to estimate the value of a square root of a number not in the table above. You can do this by using the numbers in the table above, e.g. $\sqrt{70}$ is greater than 8 and less than 9.

Cubes

The **cube** of a number is the number multiplied by itself and then by itself again. In other words it is the number multiplied by the number multiplied by the number. For example, the cube of 5, written 5^3, is equal to:

$$5 \times 5 \times 5 = 25 \times 5$$
$$= 125$$

You are expected to know the cubes of the numbers 1, 2, 3, 4, 5 and 10.

Number	Cube of number
1	1
2	8
3	27
4	64
5	125
10	1000

The numbers 1, 8, 27, 64, 125 and 1000 are all **cube numbers**.

If you cube a negative number the answer is always negative, e.g.
$-3 \times -3 \times -3 = -27$ or $(-3)^3 = -27$.

Cube roots

The **cube root** of a number, written $\sqrt[3]{}$, is the value which when cubed gives the number itself. Finding the cube root of a number is the reverse of cubing a number.

A cube root only has one value; this has the same sign as the number.

Number	Cube root of number
1	1
8	2
27	3
64	4
125	5
1000	10

EXAMINER **TIP**

Make sure you know how to use the cube root button on your calculator.

Estimating values of cube roots

In the exam, you may be asked to estimate the value of the cube root of a number not in the table above. You can do this by using the numbers in the table above, e.g. $\sqrt[3]{40}$ is greater than 3 and less than 4.

Rounding

You round a number to obtain an approximate value of the number. Sometimes you will be told the level of accuracy required, but you should also round off numbers if they are unreasonable or unmanageable, e.g. 3.77812 miles becomes 3.8 miles or 4 miles.

Decimal places

If you are asked to round a number to, for example, 1 decimal place you should follow these rules:

Look at the digit in the 2nd decimal place:

1 If the digit in the 2nd decimal place is **5 or more** you round the digit in the 1st decimal place **up** and discard all of the digits that appear after that.

Number	Number to 1 decimal place	Number to 2 decimal places
6.7245	6.7	6.72
2.154	2.2	2.15
4.38718	4.4	4.39
16.498	16.5	16.50

2 If the digit in the 2nd decimal place is **less than 5** you discard all of the digits that appear after the 1st decimal place.

Significant figures

To round a number to a given number of significant figures, you should follow these rules:

1 locate the first significant figure (the first non-zero digit)

2 count along to the number of significant figures required

3 look at the next significant figure
- if this digit is **5 or more**, round up the previous digit and discard all of the digits that appear after that
- if this digit is **less than 5**, discard this and all the digits that appear after that

4 Where necessary insert zeros to maintain the magnitude of the number.

Number	Number to 2 significant figures
1.624	1.6
3.579	3.6
0.0255	0.026
432	430
40.9	41

When a number contains zeros after the first significant figure, the zeros are also significant. For example, the number 62.0478 has the first significant figure of 6. The zero after the decimal point is taking up a significant place and cannot be ignored. This number would be 62.0 to 3 s.f. and 62.05 to 4 s.f.

Appropriate degree of accuracy

Sometimes, questions will ask you to give answers to an 'appropriate degree of accuracy' or a 'suitable degree of accuracy'. As a general rule, 2 or 3 significant figures are usually acceptable, but you should look at the context of the question. For example:

- angles – give your answer to 1 decimal place or to the nearest integer

- distance between two towns – nearest kilometre or, for longer distances, nearest 10 kilometres.

If your data in a calculation is given to 3 significant figures, then it is sensible to give your answer to 2 significant figures.

If you are in doubt, give the same answer as you would give if you were talking to someone, e.g.

Reminder
An integer is a whole number.

Question	Suitable answer	Unsuitable answer
How far is it from Castleford to Rotherham?	35 miles	34.81 miles
How old are you?	15 years	15 years, 1 month and 12 days

Standard (index) form

In Module 5 you will be expected to know and use standard form notation, and also to interpret standard form from your calculator display.

The steps to convert a number into standard form are:

Step	Example 1	Example 2	Example 3
1 Write down the number.	381.4	0.0175	6120
2 Put in the decimal point to make the number have a value between 1 and 10.	3.814	1.75	6.12
3 To make your new number the same value as the original number multiply by 10 or 100 or 1000 or $\frac{1}{10}$ or $\frac{1}{100}$ etc.	3.814×100	$1.75 \times \frac{1}{100}$	6.12×1000
4 Now write it out using powers of 10.	3.814×10^2	1.75×10^{-2}	6.12×10^3

> *Reminder*
> Multiplying by $\frac{1}{100}$ is the same as dividing by 100.

To convert a number in standard form back to an ordinary number, reverse the process:

Step	Example 1	Example 2	Example 3
1 Write down the standard form number.	3.814×10^2	1.75×10^{-2}	6.12×10^3
2 Write the power of 10 as 10 or 100 or 1000 or $\frac{1}{10}$ or $\frac{1}{100}$ etc.	3.814×100	$1.75 \times \frac{1}{100}$	6.12×1000
3 Write down the ordinary number.	381.4	0.0175	6120

You should make sure that you know how to put standard form numbers into your calculator using the EXP or EE buttons. Not all calculators are the same.

Percentage of a quantity

To find the percentage of a quantity, you first convert the percentage to a fraction or decimal and then multiply by the quantity.

Example 1.1

Work out 15% of £40.

Solution

$$15\% = \frac{15}{100} = 0.15$$

So 15% of £40 $= \frac{15}{100} \times 40$ Or 15% of £40 $= 0.15 \times 40$

$$= \frac{600}{100}$$ $= £6$

$$= £6$$

Finding a percentage of a quantity without a calculator

On the non-calculator paper, you will be expected to know how to calculate some common percentages without using a calculator.

It can be useful to know the following percentages as fractions and decimals.

Percentage	Equivalent decimal	Equivalent fraction
5%	0.05	$\frac{1}{20}$
10%	0.1	$\frac{1}{10}$
20%	0.2	$\frac{1}{5}$
25%	0.25	$\frac{1}{4}$
50%	0.5	$\frac{1}{2}$
75%	0.75	$\frac{3}{4}$

Using 10% to find other percentages

If you know 10% of a quantity you can find 5% by halving the value for 10%.

If you know 10% you can find multiples of 10%, e.g. 30% of a quantity = 3 × 10% of a quantity.

This is called a **build-up method**. You could use it to find 17.5% by working out 10% + 5% + 2.5%.

Finding a percentage of a quantity with a calculator

The method to find the percentage of a quantity is the same as before: write the percentage as a fraction or a decimal and multiply by the quantity, e.g.

$$37\% \text{ of } 84 = \frac{37}{100} \times 84 \text{ or } 0.37 \times 84$$

$$= 31.08 \qquad = 31.08$$

Practice questions

1 Write down all the square numbers from this list:

  2   13   16   27   49   80   91   121   132

2 Work out the value of:

  a 7^2 **b** 11^2 **c** 14^2

3 Work out the value of:

  a 9.4^2 **b** 13.1^2 **c** 0.3^2

4 Write down the value of $8^2 + \sqrt{169}$.

5 Write down the exact value of $(\sqrt{29})^2$.

6 Write down both square roots of 144.

7 Write down all the cube numbers from this list:

  16   40   64   100   125   1000

8 Work out the value of:

  a 7^3 **b** 8^3 **c** 30^3

9 Use a calculator to work out the value of:

  a 4.1^3 **b** 10.3^3 **c** 0.5^3

10 Put this list in ascending order of size:
   4.1^3 68.12   68.5 $68\frac{2}{3}$ $\sqrt[3]{314432}$

11 Write down the value of $5^3 - \sqrt[3]{64}$.

12 Write down the exact value of $(\sqrt[3]{612})^3$.

13 Carry out the following calculations, giving your answers correct to
   1 decimal place:

  a $23.5 + 56.75$ **b** 38.64×1.9
  c $132.1 \div 4.25$ **d** $224.4 - 180.63$

> **Reminder**
> Use your calculator to work out the full answer and then round your answer to 1 decimal place.

14 Carry out the following calculations, giving your answers correct to
   2 significant figures:

  a $12.3 + 87.67$ **b** 72.11×0.63
  c $89.32 \div 1.59$ **d** $321.9 - 91.23$

15 The formula for the volume of a sphere is $\frac{4}{3}\pi r^3$. Given that $r = 4$ cm,
   calculate the volume of the sphere.
   Give your answer to an appropriate degree of accuracy.

16 Carry out the following calculations, giving your answer **i** as an ordinary
   number and **ii** in standard form.

  a 53×8.1 **b** 168×0.0005 **c** $15\,000 \div 0.025$

17 Do not use a calculator for this question. Work out 17.5% of £600.

18 Do not use a calculator for this question. Work out 90% of 840 kg.

19 Work out 23% of 454 g.
Give your answer to an appropriate degree of accuracy.

20 Work out 109% of 53 litres.
Give your answer to an appropriate degree of accuracy.

2 Equivalent fractions

You will need to know how to:

● write fractions in their simplest form

● compare the size of fractions and order fractions.

Writing fractions in their simplest form

When writing out your answers, you may be asked to write fractions in their simplest form.

You can see that both of the rectangles below have the same fraction shaded.

   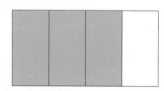

In this diagram $\frac{6}{8}$ is shaded.   In this diagram $\frac{3}{4}$ is shaded.

So $\frac{6}{8}$ is the same as $\frac{3}{4}$. They are **equivalent fractions**.

$$\frac{6}{8} = \frac{3}{4}$$

To **simplify** a fraction you find a common factor of the numerator and denominator and divide them both by that factor. This is also known as **cancelling**.

In this case, 2 is a common factor of 6 and 8.

So $\frac{6}{8} = \frac{6 \div 2}{8 \div 2}$

$= \frac{3}{4}$

> *Reminder*
> A factor is a number that will divide exactly into another number.

Simplifying fractions is completed faster if you can spot the **highest common factor** (HCF) of the numerator and denominator.

Example 2.1

Write the fraction $\dfrac{28}{36}$ in its simplest form.

Solution

The highest common factor (HCF) of 28 and 36 is 4. This means that the highest number which will divide exactly into both 28 and 36 is 4.

So $\dfrac{28}{36} = \dfrac{28 \div 4}{36 \div 4}$

$\qquad = \dfrac{7}{9}$

If you do not spot the highest common factor it is still possible to simplify the fraction:

2 is a factor of 28 and 36 as they are both even.

So $\dfrac{28}{36} = \dfrac{28 \div 2}{36 \div 2}$

$\qquad = \dfrac{14}{18}$

We can now simplify this further as 2 is also a factor of 14 and 18.

So $\dfrac{14}{18} = \dfrac{14 \div 2}{18 \div 2}$

$\qquad = \dfrac{7}{9}$

 EXAMINER TIP

It does not matter which method you use provided that you divide the numerator and denominator by the same factor.

Practice question 1

1   Write each of the following in its simplest form:

     a $\dfrac{6}{10}$ **b** $\dfrac{80}{100}$ **c** $\dfrac{6}{15}$ **d** $\dfrac{18}{24}$ **e** $\dfrac{20}{35}$ **f** $\dfrac{24}{36}$ **g** $\dfrac{90}{360}$ **h** $\dfrac{72}{108}$

Comparing the size of fractions

You may be asked to compare the size of different fractions by putting them in order of size or by looking to see which fraction is closest to a particular value.

Example 2.2

Write $\frac{1}{3}, \frac{1}{2}, \frac{1}{4}, \frac{3}{4}$ in order of size.

Solution

Common denominator method

The common denominator of 3, 2 and 4 is 12 because 12 is a multiple of 3, 2 and 4.

$$\frac{1}{3} = \frac{1 \times 4}{3 \times 4} \qquad \frac{1}{2} = \frac{1 \times 6}{2 \times 6} \qquad \frac{1}{4} = \frac{1 \times 3}{4 \times 3} \qquad \frac{3}{4} = \frac{3 \times 3}{4 \times 3}$$

$$= \frac{4}{12} \qquad\qquad = \frac{6}{12} \qquad\qquad = \frac{3}{12} \qquad\qquad = \frac{9}{12}$$

Put the twelfths in order of size. $\frac{3}{12}, \frac{4}{12}, \frac{6}{12}, \frac{9}{12}$

Now go back to the original fractions. $\frac{1}{4}, \frac{1}{3}, \frac{1}{2}, \frac{3}{4}$

Example 2.3

Which fraction is bigger, $\frac{2}{3}$ or $\frac{3}{4}$?

Solution

Method A (diagram method)

Looking at the diagrams $\frac{3}{4}$ is a larger area than $\frac{2}{3}$.

So $\frac{3}{4}$ is bigger than $\frac{2}{3}$.

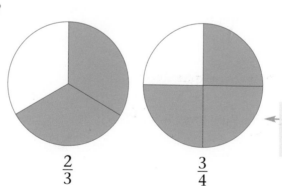

$$\frac{2}{3} \qquad\qquad \frac{3}{4}$$

EXAMINER **TIP**
Make sure that if you use this method the shapes are the same size.

Method B (decimal method)

$\frac{2}{3}$ means 2 divided by 3, $\quad \dfrac{0.666...}{3\overline{|2.000...}} \quad$ so $\frac{2}{3} = 0.666...$

$\frac{3}{4}$ means 3 divided by 4, $\quad \dfrac{0.75}{4\overline{|3.00}} \quad$ so $\frac{3}{4} = 0.75$

Looking at the decimals 0.75 is bigger than 0.666... so $\frac{3}{4}$ is bigger than $\frac{2}{3}$.

Example 2.4

Which of these fractions is closest to $\frac{1}{2}$?

$$\frac{11}{20} \qquad \frac{14}{30} \qquad \frac{29}{60}$$

Solution

First convert the fractions to have a common denominator.
The lowest common denominator is 60.

$$\frac{1}{2} = \frac{1 \times 30}{2 \times 30} \qquad \frac{11}{20} = \frac{11 \times 3}{20 \times 3} \qquad \frac{14}{30} = \frac{14 \times 2}{30 \times 2} \qquad \frac{29}{60}$$

$$= \frac{30}{60} \qquad\qquad = \frac{33}{60} \qquad\qquad = \frac{28}{60} \qquad\qquad = \frac{29}{60}$$

So $\frac{29}{60}$ is the closest fraction to $\frac{1}{2}$ $\left(\frac{30}{60}\right)$ as the difference is only $\frac{1}{60}$.

Practice questions 2

1 Put the following fractions in order of size, smallest first.

   a $\frac{2}{5}, \frac{1}{3}, \frac{1}{4}$ b $\frac{3}{4}, \frac{7}{10}, \frac{2}{3}$

   c $\frac{1}{2}, \frac{4}{9}, \frac{5}{6}$ d $\frac{3}{5}, \frac{7}{10}, \frac{13}{20}, \frac{19}{30}$

2 Which of these fractions is closest to $\frac{1}{2}$?

   $$\frac{3}{4} \qquad \frac{3}{10}$$

   Explain your answer.

3 Which of these fractions is furthest away from 1?

   $$\frac{14}{15} \qquad \frac{19}{20} \qquad \frac{24}{25}$$

   Explain your answer.

Practice exam question

1 Which of the following fractions is nearest to $\frac{1}{2}$?

   $$\frac{4}{10} \qquad \frac{9}{20} \qquad \frac{14}{30} \qquad \frac{19}{40}$$

   Show how you decide.

   [AQA (NEAB) 2001]

3 Collecting like terms

Numbers and letters in an algebraic expression are called terms. For example, $2p + 4q - 5$ is an expression that has three terms: a p-term, a q-term, and a number term.

Algebraic expressions contain letters that represent values. The letters are sometimes called variables because their values can be changed.

You may be asked to simplify an algebraic expression that contains two or more terms.

$2a$ and $3a$ are examples of **like terms** because both are terms in a.

$4a$ and $5b$ are examples of **unlike terms** because $4a$ is a term in a and $5b$ is a term in b.

$6a^2$ and $7a$ are unlike terms because $6a^2$ is a term in a^2 and $7a$ is a term in a.

You can think of simplifying expressions in the same way as adding different fruits to a shopping bag.

Here the letter a could represent apples and the letter b could represent bananas.

$2a + 3a = 5a$	2 apples + 3 apples = 5 apples
$4a + 5b$ does not simplify	4 apples + 5 bananas cannot be put any simpler
$6a - 2a = 4a$	6 apples – 2 apples = 4 apples
$5b - 9b = -4b$	5 bananas – 9 bananas = –4 bananas
	(4 bananas short)

> **Reminder**
> a is the same as $1a$. It is not necessary to write the number 1.

> **Reminder**
> $-b$ is the same as $-1b$. It is not necessary to write the number 1.

> **Reminder**
> $3 \times x$ can be written as $3x$, but not $x3$.

Example 3.1

Simplify $5p + 2q + 3p - 6q$.

Solution

Collecting the terms in p together and the terms in q together gives:

$5p + 3p + 2q - 6q = 8p - 4q$

Example 3.2

Simplify $7ab + 2a + 3a - 6a^2 - ab$.

Solution

Collecting the terms in a together and the terms in ab together gives:

$2a + 3a + 7ab - ab - 6a^2 = 5a + 6ab - 6a^2$

Practice question

1 Simplify each of the following expressions.

    a $a + b + 2a - 3b$ **b** $4c + 2d + 6c - d$

    c $5e - 2f - 6e - f$ **d** $3g + 3g - 6g - g$

    e $4h + 2h - 7h$ **f** $17k + 13m - 9n - 12m - 18k + 9n$

    g $11p + 6q - 12p + 4q + 2p - 11q$ **h** $8r + 5s - 17r - 5s$

    i $2xy + 3x + 3xy - x$ **j** $4w^2 + 3w + 2w^2 - 2w$

Practice exam questions

1 Simplify $2p + 3q - p - 4q$.   [AQA 2003]

2 Simplify $4x + 3x + 7y - 2x + 3y$.   [AQA (NEAB) 2000]

3 Simplify $9p + 4q + 6p - 7q$.   [AQA (NEAB) 2001]

4 Simplify $8a + ab - a + 2b + 3ab$.   [AQA (NEAB) 2000]

4 Rules of indices

You may be asked to simplify an algebraic expression that contains powers or indices.

An index or power tells you how many times a base is multiplied by itself, e.g. $2^5 = 2 \times 2 \times 2 \times 2 \times 2$.

Index or power

2^5

Base

You will have already covered work on index laws involving numbers in Module 3.

> **Reminder**
>
> Indices is the plural of index.

> **Reminder**
>
> An index is the power of a number or algebraic expression, e.g. the index of 2^5 is 5 and the index of p^4 is 4.

Remember the rules of indices, you treat terms with indices exactly the same way you treat numbers with indices.

To **multiply** the same value with different powers the rule is **add the indices**, e.g.

$$m^5 \times m^4 = m^9$$

To **divide** the same value with different powers the rule is **subtract the indices**, e.g.

$$\frac{d^8}{d^3} = d^5$$

To **raise** a value to a power to another **power** the rule is to **multiply the indices**, e.g.

$$(r^6)^7 = r^{42}$$

EXAMINER **TIP**

It is worth learning the rules of indices, as using the rules will be quicker than working out the answer from scratch.

Example 4.1

Simplify:

a $x^2 \times x^7$ **b** $\dfrac{y^6}{y^4}$ **c** $(z^2)^9$

Solution

a Adding the indices gives $x^2 \times x^7 = x^9$.

b Subtracting the indices gives $\dfrac{y^6}{y^4} = y^2$.

c Multiplying the indices gives $(z^2)^9 = z^{18}$.

Example 4.2

Simplify:

a $(3x^2 y) \times (4x^3 y^2)$ **b** $\dfrac{12p^5 q^2 r}{3p^3 q^2}$ **c** $\dfrac{(6st^4 u^2) \times (5s^2 t u^3)}{3s^4 t^3 u^5}$

Solution

Consider each of the like terms separately.

a $3 \times 4 = 12$; $x^2 \times x^3 = x^5$; $y \times y^2 = y^3$

So $(3x^2 y) \times (4x^3 y^2) = 12 \times x^5 \times y^3$
$$= 12x^5 y^3$$

b $12 \div 3 = 4$; $p^5 \div p^3 = p^2$; $q^2 \div q^2 = 1$

r is on its own on the top, so it stays as r.

So $\dfrac{12p^5 q^2 r}{3p^3 q^2} = 4 \times p^2 \times 1 \times r$
$$= 4p^2 r$$

Reminder

Numbers (and letter symbols) can be multiplied in any order. The result is still the same, e.g.
$(3x^2 y) \times (4x^3 y^2) = (3 \times 4) \times (x^2 \times x^3) \times (y \times y^2)$.

c First consider the top of the fraction

$6 \times 5 = 30; \quad s \times s^2 = s^3; \quad t^4 \times t = t^5; \quad u^2 \times u^3 = u^5$

So $(6st^4u^2) \times (5s^2tu^3) = 30 \times s^3 \times t^5 \times u^5$

$\qquad\qquad\qquad\qquad\qquad = 30s^3t^5u^5$

You can rewrite the fraction as $\dfrac{30s^3t^5u^5}{3s^4t^3u^5}$.

Now consider the simplified fraction.

$30 \div 3 = 10; \quad s^3 \div s^4 = s^{-1}$

$t^5 \div t^3 = t^2; \quad u^5 \div u^5 = 1$

So $\dfrac{30s^3t^5u^5}{3s^4t^3u^5} = 10 \times \dfrac{1}{s} \times t^2 \times 1$

$\qquad\qquad = \dfrac{10t^2}{s}$

> **Reminder**
>
> $s^{-1} = \dfrac{1}{s} \qquad 5^{-1} = \dfrac{1}{5}$
>
> $s^{-2} = \dfrac{1}{s^2} \qquad 5^{-2} = \dfrac{1}{5^2}$
>
> $s^{-3} = \dfrac{1}{s^3} \qquad 5^{-3} = \dfrac{1}{5^3}$

Practice question

1 Simplify the following expressions:

a $a^7 \times a^3$ 　　　　**b** $b^{13} \times b^2$

c $c \times c^4$ 　　　　**d** $\dfrac{d^7}{d^5}$

e $\dfrac{e^4}{e^4}$ 　　　　**f** $\dfrac{f^3}{f^5}$

g $(g^2)^3$ 　　　　**h** $(h^8)^2$

i $\dfrac{i^4 \times i^2}{i^3}$ 　　　　**j** $2j^3k^2 \times 4jk^3$

k $7l^6m^5 \times 3lp^2$ 　　**l** $\dfrac{4q^4r}{2q^3r}$

m $\dfrac{6st^3 \times 2t^2}{3st}$ 　　**n** $\dfrac{8uv^2 \times 4u^3v}{2u^5 \times 3uv^3}$

o $\dfrac{9wxyz \times 4w^2z^3}{2xy \times 3w}$

Practice exam questions

1 Simplify, giving your answer in index form:

    a $\dfrac{a^8}{a^4}$ **b** $(a^2)^3$ [AQA (NEAB) 2001]

2 Simplify:

    a $t^6 \times t^3$ **b** $t^6 \div t^3$ **c** $\dfrac{t^3 \times t^3}{t^2}$ [AQA (NEAB) 2002]

3 Simplify fully $\dfrac{2a^3b^2 \times 6a^4b^2}{4ab^3}$.   [AQA (NEAB) 2001]

4 **a**   Simplify the following expressions.

        i $a^5 \times a^3$ **ii** $a^5 \div a^3$ **iii** $(a^5)^3$

    b **i**   Which of these expressions is negative when $a = -1$?

           $a^5 \times a^3$ $a^5 \div a^3$ $(a^5)^3$

       ii   Which of these expressions has the greatest value when $a = 0.1$?

           $a^5 \times a^3$ $a^5 \div a^3$ $(a^5)^3$ [AQA (NEAB) 2000]

5 Expanding brackets

Expanding brackets means to remove brackets or **multiply out brackets**. The terms inside a bracket are each multiplied by the number or term in front of the bracket.

Expressions involving one variable

The terms will often consist of a variable (e.g. x, $2n$) and a constant (e.g. 13, −5) in the bracket.

Example 5.1

Expand:

a $4(2 + 7)$ b $3(x + 5)$ c $4(x - 2)$ d $7(2x + 3)$

Solution

Method A (expanding the brackets)
The terms inside the bracket are each multiplied by the number outside the bracket so:

a $4(2 + 7) = (4 \times 2) + (4 \times 7)$
                 $= 8 + 28$
                 $= 36$

b $3(x + 5) = (3 \times x) + (3 \times 5)$
                   $= 3x + 15$

c $4(x - 2) = (4 \times x) + (4 \times -2)$
                $= 4x - 8$

d $7(2x + 3) = (7 \times 2x) + (7 \times 3)$
                    $= 14x + 21$

Method B (grid or box method)

Write each expression in a multiplication box:

a

×	2	+7
4	8	+28

Then $4(2 + 7) = 8 + 28$
$\qquad\qquad\quad = 36$

b

×	x	+5
3	$3x$	+15

Then $3(x + 5) = 3x + 15$.

c

×	x	−2
4	$4x$	−8

Then $4(x − 2) = 4x − 8$.

d

×	$2x$	+3
7	$14x$	+21

Then $7(2x + 3) = 14x + 21$.

Example 5.2

Expand:

a $-3(x + 5)$ **b** $-4(x - 2)$ **c** $-2(3x + 7)$ **d** $-5(2 - 3x)$

Solution

a $-3(x + 5) = (-3 \times x) + (-3 \times 5)$
$\qquad\qquad\quad = -3x - 15$

b $-4(x - 2) = (-4 \times x) + (-4 \times -2)$
$\qquad\qquad\quad = -4x + 8$

c $-2(3x + 7) = (-2 \times 3x) + (-2 \times 7)$
$\qquad\qquad\qquad = -6x - 14$

d Write the expression $-5(2 - 3x)$ in a multiplication table.

×	2	−3x
−5	−10	+15x

Then $-5(2 - 3x) = -10 + 15x$.

> **Reminder**
> A negative number multiplied by a positive number will always have a negative answer. So $-3 \times +5 = -15$.

> **Reminder**
> Two negative numbers multiplied together will always have a positive answer. So $-4 \times -2 = +8$.

Example 5.3

Expand:

a $x(x + 5)$ b $x(x - 4)$ c $x(2x + 7)$ d $-x(x + 4)$ e $-x(2x - 9)$

Solution

a $x(x + 5) = x^2 + 5x$ b $x(x - 4) = x^2 - 4x$

c $x(2x + 7) = 2x^2 + 7x$ d $-x(x + 4) = -x^2 - 4x$

e Write the expression $-x(2x - 9)$ in a multiplication table.

×	$2x$	-9
$-x$	$-2x^2$	$+9x$

Then $-x(2x - 9) = -2x^2 + 9x$.

Reminder
When two of the same variables are multiplied together the result is a variable raised to the power 2. So $x \times 2x = 2x^2$.

Expressions with more than one variable

There is no reason why other letters cannot be used but the same methods apply.

Example 5.4

Expand:

a $p(p + 2q)$ b $r(t - 3r)$ c $s(2s - 5)$ d $3s(2 + 3s)$ e $x(y - 2x + 3z)$

Solution

a Using the box method write $p(p + 2q)$ in a multiplication table.

×	p	$+2q$
p	p^2	$+2pq$

Then $p(p + 2q) = p^2 + 2pq$.

Reminder
$2pq$ means $2 \times p \times q$. We often write the product of letters in alphabetical order but $2pq$ is exactly the same as $2qp$.

b $r(t - 3r) = rt - 3r^2$ c $s(2s - 5) = 2s^2 - 5s$

d $3s(2 + 3s) = 6s + 9s^2$ e $x(y - 2x + 3z) = xy - 2x^2 + 3xz$

Practice question 1

1 Expand:

  a $3x(2x + 4y - z)$ b $t(2t + 3r)$ c $4p(q - 2p)$

  d $-2x(3x + y - z)$ e $x(x^2 + 3x - 4)$

Expanding brackets and simplifying terms

Questions sometimes involve expanding brackets and simplifying the terms by combining the 'like' terms.

Example 5.5

Expand and simplify:

a $3(x - 6) + 2x$ b $5x - 2(x + 3)$ c $4(x + 5) - 8$

d $5p(2p + 3) - 4p$ e $3y(2y + 5) - 7(y + 2)$ f $(7h + 3t) - (h - t)$

Solution

a **Step 1**
Expand the brackets: $3(x - 6) + 2x = 3x - 18 + 2x$
$3x$ and $2x$ are called like terms because they contain the same variable x.
Step 2
You can combine the like terms: $3x + 2x = 5x$, which simplifies the answer, giving $3(x - 6) + 2x = 5x - 18$.

b **Step 1**
Expand the brackets: $5x - 2(x + 3) = 5x - 2x - 6$
$5x$ and $-2x$ are called like terms because they contain the same variable x.
Step 2
You can combine the like terms: $5x - 2x = 3x$, which simplifies the answer, giving $5x - 2(x + 3) = 3x - 6$.

c Expand the brackets: $4(x + 5) - 8 = 4x + 20 - 8$
Now we simplify by combining like terms: $20 - 8 = 12$
This gives $4(x + 5) - 8 = 4x + 12$.

d Expand the brackets: $5p(2p + 3) - 4p = 10p^2 + 15p - 4p$
Here the p^2-term is different to the p-terms. Only the p-terms are like terms and can be combined: $15p - 4p = 11p$
This gives $5p(2p + 3) - 4p = 10p^2 + 11p$.

e $3y(2y + 5) - 7(y + 2) = 6y^2 + 15y - 7y - 14$
Simplify the like terms: $15y - 7y = 8y$
This gives $3y(2y + 5) - 7(y + 2) = 6y^2 + 8y - 14$.

f A minus in front of a bracket without a number means -1 multiplied by that bracket.
$(7h + 3t) - (h - t) = 7h + 3t - h + t$
$= 6h + 4t$

> *Reminder*
> A negative number multiplied by a negative number always gives a positive answer.

Practice questions 2

Expand and simplify:

1 a $5(x - 3) + 2x$ **b** $8x - 2(x + 5)$ **c** $7(x - 5) + 6x$ **d** $2p(3 - 4p) + 5p - 7$

2 a $5g(2g + 3) - 6g$ **b** $3y(2y - 7) - 3(y + 1)$ **c** $4(x - 5) - 6(x - 3)$ **d** $2(2a + 8) + 5(3a - 7)$

Expanding brackets involving powers

Occasionally you may be asked to expand a bracket where some of the terms are raised to different powers.

Reminder
For help with terms raised to powers see Chapter 4, Rules of indices.

Example 5.6

Expand $6a^4(3a^2 - b)$.

Solution

$$6a^4(3a^2 - b) = 6a^4 \times 3a^2 - 6a^4 \times b$$
$$= 18a^6 - 6a^4b$$

Reminder
When multiplying like terms, the indices are added, e.g. $a^3 \times a = a^4$.

Practice question 3

1   Expand and simplify:

   a $6a(3a^3 - 2b)$ **b** $7x^3(2x^2 - y)$ **c** $4y^2(5 - 3y)$ **d** $(2x + 3y)5x^2$ **e** $4t(3t^3 - 2t^2 + 3t - 6)$

Expanding two brackets

To expand two brackets multiply each term in the first bracket by each term in the second bracket.

Example 5.7

Expand and simplify:

a $(x + 3)(x + 7)$ **b** $(x + 2)(x + 5)$ **c** $(x - 4)(x + 3)$ **d** $(x - 5)(x - 3)$

Reminder
The minus sign in part **c** is attached to the number 4.

Solution

a $(x + 3)(x + 7) = x(x + 7) + 3(x + 7)$
$\qquad\qquad\qquad = x^2 + 7x + 3x + 21$

Now you add the like terms $7x + 3x = 10x$ giving $(x + 3)(x + 7) = x^2 + 10x + 21$.

b $(x + 2)(x + 5) = x(x + 5) + 2(x + 5)$
$\qquad\qquad\qquad = x^2 + 5x + 2x + 10$

Now you add the like terms $5x + 2x = 7x$ giving $(x + 2)(x + 5) = x^2 + 7x + 10$.

c $(x - 4)(x + 3) = x(x + 3) + -4(x + 3)$
$\qquad\qquad\qquad = x^2 + 3x + -4x - 12$

Now you add the like terms $3x + -4x = -x$ giving $(x - 4)(x + 3) = x^2 - x - 12$.

d $(x - 5)(x - 3) = x(x - 3) + -5(x - 3)$
$\qquad\qquad\qquad = x^2 - 3x + -5x + 15$

Now you add the like terms $-3x + -5x = -8x$ giving $(x - 5)(x - 3) = x^2 - 8x + 15$.

Reminder
$-5 \times -3 = +15$

Practice question 4

1 Expand and simplify:

  a $(x + 6)(x + 7)$ **b** $(x + 4)(x + 2)$ **c** $(x - 4)(x + 6)$

  d $(x - 5)(x - 1)$ **e** $(x - 1)(x + 9)$

The grid or box method

If you consider a square of side x cm, then the area of that square would be $x \times x = x^2$ cm^2.

Now consider a rectangle whose sides are $(x + 2)$ cm and $(x + 7)$ cm, then the area of this rectangle would be $(x + 2)(x + 7)$ cm^2.

You can represent this area with the following diagram.

From the diagram the area of the four parts added together is the area of the large rectangle.

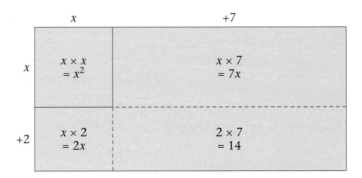

So $(x + 2)(x + 7) = x^2 + 7x + 2x + 14$

Then $(x + 2)(x + 7) = x^2 + 9x + 14$

This is used in a grid to help you work out the four products when multiplying brackets.

> **Reminder**
> A product is the result when two or more terms are multiplied together.

Example 5.8

Use the grid method to expand the following pairs of brackets.

a $(q + 3)(q + 5)$ **b** $(r - 4)(r - 5)$ **c** $(2x + 1)(3x - 2)$

Solution

a $(q + 3)(q + 5)$

×	q	+3
q	q^2	+3q
+5	+5q	+15

Like terms

Then $(q + 3)(q + 5) = q^2 + 3q + 5q + 15$.

Adding the like terms gives $(q + 3)(q + 5) = q^2 + 8q + 15$.

b $(r-4)(r-5)$

×	r	-4
r	r^2	$-4r$
-5	$-5r$	$+20$

Like terms

Then $(r-4)(r-5) = r^2 - 4r - 5r + 20$

so $(r-4)(r-5) = r^2 - 9r + 20$

c $(2x+1)(3x-2)$

×	$2x$	$+1$
$3x$	$6x^2$	$+3x$
-2	$-4x$	-2

Like terms

Then $(2x+1)(3x-2) = 6x^2 + 3x - 4x - 2.$

Combining the like terms gives $(2x+1)(3x-2) = 6x^2 - x - 2.$

Practice question 5

1 Use the grid method to expand the following pairs of brackets.

    a $(t+2)(t+8)$ **b** $(u-4)(u-3)$ **c** $(3x+2)(2x-1)$

The smiley face method

This is a different method of expanding two brackets.

You draw in pencil links connecting both terms in the first bracket to both terms in the second bracket.

The links look a bit like a smiley face. They show the terms that are to be multiplied.

Step 1 $x \times x = x^2$

Step 2 $x \times +6 = +6x$

Step 3 $+5 \times x = +5x$

Step 4 $+5 \times +6 = +30$

You now add the like terms: $+6x + 5x = +11x.$

So $(x+5)(x+6) = x^2 + 11x + 30.$

Example 5.9

Expand and simplify $(x + 4)(x + 9)$.

Solution

First draw the links between terms. $(x + 4)(x + 9)$

Step 1 $x \times x = x^2$ **Step 2** $x \times +9 = +9x$

Step 3 $+4 \times x = +4x$ **Step 4** $+4 \times +9 = +36$

You now add the like terms: $+9x + 4x = +13x$.
So $(x + 4)(x + 9) = x^2 + 13x + 36$.

Example 5.10

Expand and simplify $(x + 2)(x - 3)$.

Solution

$(x + 2)(x - 3)$

Step 1 $x \times x = x^2$ **Step 2** $x \times -3 = -3x$

Step 3 $+2 \times x = +2x$ **Step 4** $+2 \times -3 = -6$

Now add the like terms $-3x + 2x = -x$.
So $(x + 2)(x - 3) = x^2 - x - 6$.

Example 5.11

Expand and simplify $(x - 5)(x - 4)$.

Solution

$(x - 5)(x - 4)$

Step 1 $x \times x = x^2$ **Step 2** $x \times -4 = -4x$

Step 3 $-5 \times x = -5x$ **Step 4** $-5 \times -4 = +20$

Now add the like terms $-4x + -5x = -9x$.
Write out the answer $(x - 5)(x - 4) = x^2 - 9x + 20$.

Example 5.12

Expand and simplify $(4x + 5)(2x - 3)$.

Solution

$$(4x + 5)(2x - 3) = 8x^2 - 12x + 10x - 15$$
$$= 8x^2 - 2x - 15$$

Squaring brackets

Squaring a bracket involves multiplying the bracket by itself. It is best to write the bracket out twice side by side and then expand as above.

Example 5.13

Expand and simplify $(x + 3)^2$.

Solution

$$(x + 3)^2 = (x + 3)(x + 3)$$
$$= x^2 + 3x + 3x + 9$$
$$= x^2 + 6x + 9$$

Practice questions 6

Expand and simplify the following.

1   **a** $(x - 4)(x + 3)$ **b** $(x - 5)(x + 4)$

2   **a** $(2x - 4)(x + 1)$ **b** $(2x + 3)(4x + 5)$ **c** $(3p - 4)(5p - 7)$

3   **a** $(x + 5)^2$ **b** $(x + 9)^2$ **c** $(x - 3)^2$ **d** $(x - 6)^2$

4   **a** $(2x + 1)^2$ **b** $(3x + 2)^2$ **c** $(2x - 4)^2$ **d** $(2x - 3)^2$

5   **a** $(x + y)^2$ **b** $(2x - y)^2$ **c** $(3x + 2y)^2$ **d** $(3x - 4y)^2$

Practice exam questions

1   Multiply out $3(2c - 5)$.   [AQA (SEG) 2000]

2   Expand and simplify:
    a $4(x + 3) + 6(x + 1)$ **b** $(3x + 2y) - (x - y)$ [AQA 2002]

3   Expand:
    a $2x(3 + 2y)$ **b** $4a^3(a^2 - b)$ [AQA (NEAB) 1999]

4   Expand and simplify:
    a $5x - 4(2x - 5)$ **b** $(3x + 2y)(4x - y)$ [AQA (NEAB) 2002]

5   Expand and simplify $(2x + 3y)^2$.   [AQA (NEAB) 2000]

6 Changing the subject of a formula

A **formula** is an algebraic rule. It always has an equals sign in it.
The formula for the area A of a circle of radius r is $A = \pi r^2$.
A is the **subject** of the formula.
The formula can be rearranged to make r the subject of the formula.

$$A = \pi r^2$$

Divide both sides by π:
$$\frac{A}{\pi} = r^2$$

Take the square root of both sides: $\sqrt{\dfrac{A}{\pi}} = r$

Sometimes it reads better if the subject is on the left-hand side of the equation:

$$r = \sqrt{\frac{A}{\pi}}$$

EXAMINER **TIP**

Normally when taking the square root there can be a positive and a negative root. However, a radius cannot be negative, so you only have the positive root.

Example 6.1

$v = u + at$

Rearrange the formula to make:

a u the subject   **b** t the subject.

Solution

a Write out the given formula:

$v = u + at$

To make u the subject of the formula you need to take the at term away from both sides of the equation:

$$v = u + at$$
$$v - at = u + at - at$$

Since $at - at = 0$, you can remove these from the right-hand side:

$v - at = u$ or $u = v - at$

b Write out the given formula:

$v = u + at$

To make t the subject of the formula you need to perform two steps.

Step 1
Take the u term away from both sides of the equation:

$v - u = u - u + at$

Since $u - u = 0$, you can remove these from the right-hand side (RHS):

$v - u = at$

Step 2
Divide both sides of the equation by a:

$$\frac{v - u}{a} = \frac{at}{a}$$

Reminder

Letters are just like numbers: $at - at = 0$ in the same way that $(3 \times 2) - (3 \times 2) = 0$.

The *a* terms will cancel on the RHS of the equation:

$$\frac{v-u}{a} = \frac{\not{a}t}{\not{a}}$$

$$\frac{v-u}{a} = t \text{ or } t = \frac{v-u}{a}$$

Example 6.2

Make *x* the subject of the formula.
$y = 3x - 5$

Solution

Write out the given formula:
$y = 3x - 5$

Step 1
Add 5 to both sides of the equation:
$y + 5 = 3x - 5 + 5$

Since $-5 + 5 = 0$, you can remove these from the right-hand side:
$y + 5 = 3x$

Step 2
Divide both sides of the equation by 3:
$$\frac{y+5}{3} = x \text{ or } x = \frac{y+5}{3}$$

This can also be written as $x = \frac{1}{3}y + \frac{5}{3}$.

Example 6.3

Rearrange the formula to make *t* the subject.

$P = s + \sqrt{t}$

Solution

Write out the given formula:
$P = s + \sqrt{t}$

Step 1
Subtract *s* from both sides of the equation:
$P - s = s + \sqrt{t} - s$

Since $s - s = 0$, you can remove these from the right-hand side (RHS):
$P - s = \not{s} + \sqrt{t} - \not{s}$
$P - s = \sqrt{t}$

Step 2
Square both sides:
$(P - s)^2 = t \text{ or } t = (P - s)^2$

> *Reminder*
> When you square you have to insert brackets as you are squaring the whole of the left-hand side.

Example 6.4

Rearrange this equation to make x the subject.
$3(y - x) = 2x + 7$

Solution

Write out the given equation:
$3(y - x) = 2x + 7$

In this equation, x (which is to be the subject) appears on both sides of the equation.
It is necessary to expand the brackets on the left-hand side (LHS) and move the x-terms to the right-hand side (RHS).

Step 1
Expand the brackets on the LHS:
$3y - 3x = 2x + 7$

Step 2
Add $3x$ to both sides of the equation:
$3y - 3x + 3x = 2x + 7 + 3x$

Since $-3x + 3x = 0$, you can remove these from the LHS:
$3y = 2x + 7 + 3x$

You can also combine the x-terms on the RHS:
$3y = 5x + 7$

Step 3
Subtract 7 from both sides of the equation:
$3y - 7 = 5x + 7 - 7$

Since $7 - 7 = 0$, you can remove these from the RHS:
$3y - 7 = 5x$

Step 4
Divide both sides by 5:
$$\frac{3y - 7}{5} = \frac{5x}{5}$$

The 5s cancel on the RHS:
$$\frac{3y - 7}{5} = \frac{\cancel{5}x}{\cancel{5}}$$

$$\frac{3y - 7}{5} = x \text{ or } x = \frac{3y - 7}{5}$$

Example 6.5

Make q the subject of the formula.

$$\frac{5(q+4)}{6} = p$$

Solution

Write out the given formula:

$$\frac{5(q+4)}{6} = p$$

In this formula the left-hand side (LHS) is a fraction.

Step 1
Multiply both sides by 6:

$$\frac{5(q+4)}{6} \times 6 = p \times 6$$

The 6s cancel on the LHS:

$$\frac{5(q+4)}{\cancel{6}} \times \cancel{6} = p \times 6$$

$$5(q+4) = 6p$$

Step 2
Expand the brackets on the LHS:

$$5q + 20 = 6p$$

Step 3
Subtract 20 from both sides:

$$5q + 20 - 20 = 6p - 20$$

Since $20 - 20 = 0$, you can remove these from the LHS:

$$5q = 6p - 20$$

Step 4
Divide both sides by 5:

$$\frac{5q}{5} = \frac{6p - 20}{5}$$

The 5s cancel on the LHS:

$$\frac{\cancel{5}q}{\cancel{5}} = \frac{6p - 20}{5}$$

$$q = \frac{6p - 20}{5}$$

Practice questions

1 Make k the subject of $8k - m = 2n$.

2 Rearrange this equation to make x the subject.
   $y = 3x - 10$

3 **a** Make s the subject of $v^2 = u^2 + 2as$.
   b Make a the subject of $v^2 = u^2 + 2as$.
   c Make u the subject of $v^2 = u^2 + 2as$.

4 Rearrange the equation $3x - 2y = 4$ into the form $y = mx + c$.

5 Make b the subject of $b^2 + c = 15$.

6 Make m the subject of $n = \dfrac{3(m-2)}{5}$.

Practice exam questions

1 Make x the subject of the formula.

   $y = 5x - 6$ [AQA (NEAB) 2001]

2 Make y the subject of the formula.

   $x = 40 - 8y$ [AQA (NEAB) 1999]

3 Rearrange the following formula to make x the subject.

   $y = mx + c$ [AQA (NEAB) 1999]

4 Make x the subject of the formula.

   $w = x^2 + y$ [AQA 2003]

5 Make q the subject of the formula.

   $p = q^2 + r$ [AQA (NEAB) 2001]

6 Make t the subject of the formula.

   $W = \dfrac{5t + 3}{4}$ [AQA (NEAB) 1998]

7 Make x the subject of $x^2 + k = 16$. [AQA 2003]

8 Make p the subject of the formula.

   $\dfrac{4(p+3)}{7} = r$ [AQA (NEAB) 2000]

7 Substitution into a formula

You will be given algebraic expressions and then be asked to **substitute** positive and negative numbers in order to find the value of the expression.

Questions on the non-calculator paper will usually involve straightforward numbers such as integers or small fractions. Questions on the calculator paper may involve more complex calculations and will be testing your ability to use a calculator effectively.

Questions may involve substituting numbers into a formula from another part of the course, such as a perimeter, an area or a volume formula. You may also have to use a formula from another subject, but in this case you will always be given the formula to use.

> **Reminder**
> Don't forget the order of operations. BODMAS – Brackets, Order, Division, Multiplication, Addition, Subtraction.

Example 7.1

You are given that $x = -2$, $y = 3$ and $z = \frac{1}{2}$.

Work out the value of:

a $2x^2 + 3$ **b** $\dfrac{2y - x}{z}$ **c** $3x - y + 6z$ **d** $x^2 + x - 1$

Solution

It is always sensible to substitute the numbers in first (using brackets because of the BODMAS rules) before trying to work anything out.

a Substituting $x = -2$ into $2x^2 + 3$ gives
$$2(-2)^2 + 3 = 2 \times 4 + 3$$
$$= 8 + 3$$
$$= 11$$

> **Reminder**
> $(-2)^2 = -2 \times -2 = 4$

b Substituting $x = -2$, $y = 3$ and $z = \frac{1}{2}$ into $\dfrac{2y - x}{z}$ gives

$$\frac{2(3) - (-2)}{\frac{1}{2}} = \frac{6 + 2}{\frac{1}{2}}$$
$$= \frac{8}{\frac{1}{2}}$$
$$= 8 \div \tfrac{1}{2}$$
$$= 8 \times 2$$
$$= 16$$

> **Reminder**
> $-(-2) = +2$

> **Reminder**
> Dividing by a fraction is the same as multiplying by its reciprocal. The reciprocal of $\frac{1}{2}$ is 2. So $8 \div \frac{1}{2} = 8 \times 2$.

> **EXAMINER TIP**
> There are 16 halves in 8. A common mistake is to divide by 2 instead of dividing by $\frac{1}{2}$.

c Substituting $x = -2$, $y = 3$ and $z = \frac{1}{2}$ into $3x - y + 6z$ gives

$$3 \times (-2) - (3) + 6 \times (\tfrac{1}{2}) = -6 - 3 + 3$$
$$= -6$$

d Substituting $x = -2$ into $x^2 + x - 1$ gives
$$(-2)^2 + (-2) - 1 = 4 - 2 - 1$$
$$= 1$$

Example 7.2

Using given formulae

a You are given that $V = IR$. Find the value of V when $I = 1.75$ and $R = 24$.

b You are given that $\sigma = \dfrac{F}{A}$. Find the value of σ when $F = 1500$ and $A = 3.5$.

c You are given that $A = \pi r^2$. Find the value of A when $r = 5$ and $\pi = 3.14$.

d You are given that $E = mc^2$. Find the value of E when $m = 0.2$ and $c = 300\ 000\ 000$.

e You are given that $v = u + at$. Find the value of t when $v = 20$, $u = 6$ and $a = 2$.

f You are given that $p = 5 + \sqrt{r}$. Find the value of p when $r = 24$.

g You are given that $q = \sqrt[3]{s} + 2t$. Find the value of q when $s = 29$ and $t = 5$.

h You are given that $M = 3^h \times h^3$. Find the value of M when $h = 2$.

Solution

a Substituting $I = 1.75$ and $R = 24$ into $V = IR$ gives
$V = 1.75 \times 24$
$V = 42$

> **Reminder**
> IR means $I \times R$.

b Substituting $F = 1500$ and $A = 3.5$ into $\sigma = \dfrac{F}{A}$ gives

$\sigma = \dfrac{1500}{3.5}$

$\sigma = 429$ (to 3 significant figures)

c Substituting $r = 5$ into $A = \pi r^2$ and taking $\pi = 3.14$ gives
$A = 3.14 \times (5)^2$
$A = 78.5$

d Substituting $m = 0.2$ and $c = 300\ 000\ 000$ into $E = mc^2$ gives
$E = 0.2 \times 300\ 000\ 000^2$
$E = 18\ 000\ 000\ 000\ 000\ 000$

When using large numbers, you can use standard form notation. Some questions might ask you to write answers in standard form. This would give
$E = 0.2 \times (3 \times 10^8)^2$
$E = 1.8 \times 10^{16}$

e Substituting $v = 20$, $u = 6$ and $a = 2$ into $v = u + at$ gives
$20 = 6 + 2t$

Subtracting 6 from both sides gives
$14 = 2t$

Dividing both sides by 2 gives
$7 = t$

f Substituting $r = 24$ into $p = 5 + \sqrt{r}$ gives

$p = 5 + \sqrt{24}$

$p = 5 + 4.899$

$p = 9.899$ (3 d.p.)

g Substituting $s = 29$ and $t = 5$ into $q = \sqrt[3]{s} + 2t$ gives

$q = \sqrt[3]{29} + 2 \times 5$

$q = 3.072 + 10$

$q = 13.072$ (3 d.p.)

h Substituting $h = 2$ into $M = 3^h \times h^3$ gives

$M = 3^2 \times 2^3$

$M = 9 \times 8$

$M = 72$

Practice questions 1

1   Given that $x = 3$, $y = -2$ and $z = -\dfrac{1}{2}$, find the value of:

    a $x + y + z$

    b xyz

    c $(x + y)^2$

    d $\dfrac{(x + y)^2}{z}$

2   Given that $p = 2l + 2w$:

    a find the value of p when $l = 6.1$ and $w = 3.9$

    b find the value of l when $p = 30$ and $w = 4$.

3   Given that $P = I^2 R$:

    a find the value of P when $I = 0.25$ and $R = 17.8$

    b find the value of R when $I = 3.7$ and $P = 60$.

Function notation

You may be given questions which use **function notation**, such as $f(x) = 3x + 4$. This is a different way of writing $y = 3x + 4$.

Here is how the same question can be presented in two different ways.

Example 7.3

$y = 3x + 4$

Find the value of y when $x = 5$.

Solution

Substituting the value $x = 5$ into the equation gives:

$y = 3(5) + 4$

$y = 15 + 4$

$y = 19$

Example 7.4

$f(x) = 3x + 4$

Find the value of $f(5)$.

Solution

Substituting the value $x = 5$ into the function gives:
$f(5) = 3(5) + 4$
$f(5) = 15 + 4$
$f(5) = 19$

Practice question 2

1 Find the value of $f(1)$, $f(-1)$ and $f(3)$ for each of the following functions.

    a $f(x) = x^2$
    b $f(x) = 2x - 4$
    c $f(x) = 2x^2 + 5$
    d $f(x) = \dfrac{1}{x}$

Practice exam questions

1 Find the value of $3x + 4y$ when $x = 6$ and $y = -3$.   [AQA 2003]

2 Using $p = 18.8$, $q = 37.2$, $r = 0.4$, work out:

    a $p + \dfrac{q}{r}$

    b $\dfrac{p + q}{r}$ [AQA (NEAB) 2001]

3 You are given the formula $v = u + at$.

    Work out the value of v when $u = 20$, $a = -6$ and $t = \dfrac{9}{5}$.   [AQA (SEG) 1998]

4 Find the value of $3x + y^2$ when $x = -2$ and $y = 3$.   [AQA (SEG) 2000]

5 $d = 4e + 5h^2$
    Calculate the value of d when $e = 6.7$ and $h = 3$.   [AQA (NEAB) 2002]

6 s is given by the formula $s = ut + \frac{1}{2}at^2$.
    Find the value of s when $u = 2.8$, $t = 2$ and $a = -1.7$.   [AQA (NEAB) 2000]

7 Calculate the value of $2^a \times a^2$, when $a = -2$.   [AQA (SEG) 1999]

8 a

Southern Rental
Van Hire Charges
£24 per day plus 12 pence per mile

    i Anita hires a van for one day.   **ii** John hires the van for two days.
        She drives 68 miles.   The total hire charge is £66.
        How much is the hire charge?   How many miles did he drive?

Cars can also be hired.
When a car is hired for d days and driven m miles, the hire charge,
C pounds, is calculated by using the formula
$C = 0.06(300d + m)$

  b Use the formula to calculate the cost of hiring a car for 7 days and
      driving 458 miles.   [AQA (SEG) 1999]

9 You are given the formula $P = \dfrac{V^2}{R}$.

    Work out the value of P when $V = 3.85$ and $R = \dfrac{8}{5}$.   [AQA (SEG) 1999]

10 Work out the value of $x^2 - x - 12$ when $x = 4$.   [AQA (NEAB) 2001]

11 Use your calculator to work out $\sqrt{15 - \pi}$.   [AQA (NEAB) 1998]

8 Number patterns and sequences

Number sequences

Here are some number sequences that you should already know.

Odd numbers	1, 3, 5, 7, 9, ...
Even numbers	2, 4, 6, 8, 10, ...
Positive integers	1, 2, 3, 4, 5, ...
Negative integers	–1, –2, –3, –4, –5, ...
Square numbers	1, 4, 9, 16, 25, ...
Cube numbers	1, 8, 27, 64, 125, ...
Prime numbers	2, 3, 5, 7, 11, ...
Powers of 2	$2^0, 2^1, 2^2, 2^3, 2^4, 2^5,$ ...
	1, 2, 4, 8, 16, 32, ...
Powers of 10	$10^0, 10^1, 10^2, 10^3, 10^4,$ ...
	1, 10, 100, 1000, 10 000, ...
Triangular numbers (also called triangle numbers)	1, 3, 6, 10, 15, ...

> **Reminder**
> An integer is a whole number. It can be positive or negative.

> **Reminder**
> A prime number is a number with only two factors, itself and 1.

You will have to be able to work out the next number (term) in a sequence. The terms in a sequence will often increase or decrease by a fixed amount, for example in the sequence 3, 5, 7, 9, ... the terms increase by 2 each time. Sometimes the difference between the terms will increase or decrease in even steps, for example in the sequence 3, 5, 8, 12, ... the difference between the terms increases by 1 each time starting with a difference of 2.

Example 8.1

Write down the next two terms in the sequence.
16, 14, 11, 7, ..., ...

Solution

Look at the differences between terms

16   14   11   7   ...   ...
  Subtract 2   Subtract 3   Subtract 4

The rule here is to subtract 1 more each time.
Continuing the sequence gives:

16   14   11   7   2   –4
  Subtract 2   Subtract 3   Subtract 4   Subtract 5   Subtract 6

> **EXAMINER TIP**
> At Intermediate tier the questions will go into negative numbers.

Patterns from diagrams

You will also need to form patterns from stick diagrams.

Example 8.2

How many sticks are needed for diagram 5?

| Diagram 1 | Diagram 2 | Diagram 3 |
| 4 sticks | 7 sticks | 10 sticks |

Solution

Look at the differences again.

$$4 \xrightarrow{+3} 7 \xrightarrow{+3} 10$$

The rule here is to add 3 each time.
Diagram 4 will have $10 + 3 = 13$ sticks.
Diagram 5 will have $13 + 3 = 16$ sticks.

Alternatively, you can draw out the diagram and count the sticks.

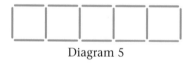

Diagram 5

16 sticks

Patterns from equations

You may be given a number pattern using a series of equations. The question will usually ask you to complete the next line of the pattern. These questions will usually appear on the non-calculator paper as you are not expected to use a calculator to work them out. You are expected to spot the patterns.

Example 8.3

Here is a number sequence.

Line 1:　　$1 \times 4 + 2 = 2 \times 3$

Line 2:　　$2 \times 5 + 2 = 3 \times 4$

Line 3:　　$3 \times 6 + 2 = 4 \times 5$

Write down:

a the next line of the sequence

b the 10th line of the sequence.

Solution

a Looking at the pattern down the columns we can obtain the answer for Line 4.

Line 1　　1　×　4　+　2　=　2　×　3

Line 2　　2　×　5　+　2　=　3　×　4

Line 3　　3　×　6　+　2　=　4　×　5

Line 4　　4　×　7　+　2　=　5　×　6

(Add 1 down each column except the "+2" column which stays the Same.)

b Look at the pattern across the rows to obtain the answer for Line 10.

Line 1	1	×	4	+	2	=	2	×	3
Line 2	2	×	5	+	2	=	3	×	4
Line 3	3	×	6	+	2	=	4	×	5
	Line number		Line number +3		Fixed at 2		Line number +1		Line number +2

Using this pattern gives:

| Line 10 | 10 | × | 13 | + | 2 | = | 11 | × | 12 |

Example 8.4

Here is a number sequence.

$$1 \times 1 = 1$$
$$11 \times 11 = 121$$
$$111 \times 111 = 12321$$

a Write down the next line of this sequence.

b Explain why you cannot write down the 10th line of this sequence.

Solution

a $1111 \times 1111 = 1234321$

b Each time the middle digit increases by 1, but 10 is a two-digit number so it would not fit into one position.

The actual answer to $1111111111 \times 1111111111$ is 1234567900987654321.

Practice questions 1

1   Write down the next two terms in each sequence and then state the rule.

   a 19, 14, 9, ..., ...
   b 6, 1, –3, –6, ..., ...
   c 9, $6\frac{1}{2}$, 4, $1\frac{1}{2}$, ..., ...
   d 13, 8, 1, –8, ..., ...

2   Here is a pattern made of sticks.

  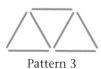 

   Pattern 1   Pattern 2   Pattern 3   Pattern 4

   How many sticks are used in Pattern 6?
   Explain your answer.

3   Write down the next two lines of this sequence.
$$(1)^2 = 1^3$$
$$(1 + 2)^2 = 1^3 + 2^3$$
$$(1 + 2 + 3)^2 = 1^3 + 2^3 + 3^3$$

4   Write down:

   a the next line of this sequence
   b the 10th line of this sequence.

$$3^2 - 2^2 = 1 \times 5$$
$$5^2 - 3^2 = 2 \times 8$$
$$7^2 - 4^2 = 3 \times 11$$

5   Write down:

   a the next line of this sequence
   b the 10th line of this sequence.

$$1 + 2 = \frac{2 \times 3}{2}$$
$$1 + 2 + 3 = \frac{3 \times 4}{2}$$
$$1 + 2 + 3 + 4 = \frac{4 \times 5}{2}$$

Linear sequences and *n*th term

A **linear sequence** is a sequence in which the next term is found by adding or subtracting a fixed amount. For example 3, 7, 11, 15, ... is a linear sequence because the next term is found by adding on 4 each time.

To find an expression for the *n*th term

The ***n*th term** is a general formula that gives the value of a term from its position number in the sequence. For example, by substituting $n = 50$ into the formula for the *n*th term you can find the value of the 50th term.

You may be asked in the exam to find an expression for the *n*th term of a sequence. You will only be asked about linear sequences.

Linear sequences always have the same difference between the terms in the sequence. This is called the **common difference**. For example, in the sequence 1, 4, 7, 10, ... the common difference is 3. Knowing the common difference will help you find the *n*th term. This is because in linear sequences the *n*th term will always be of the form ***an + c***, where *a* is the common difference.

Example 8.5

Here is a sequence of numbers.

2, 5, 8, 11, ...

a Work out an expression for the *n*th term.

b Use your answer to write down the 100th term of the sequence.

Solution

a First look at the sequence and find the common difference.

Term number (*n*)	1	2	3	4
Sequence	2	5	8	11

+3   +3   +3

The common difference is +3. This tells you that the *n*th term is $3n + c$. Now you need to look at the difference between the sequence and the value for $3n$ to find *c*.

Term number (*n*)	1	2	3	4
Sequence	2	5	8	11
3*n*	3	6	9	12

The difference between the sequence term and the 3*n* term is always −1, so this tells you that *c* must be −1.
Check the sequence values and the values using your *n*th term to see if you are correct.

Term number (*n*)	1	2	3	4
Sequence	2	5	8	11
3*n* − 1	2	5	8	11

They are the same so the *n*th term is $3n - 1$.

b The 100th term of the sequence is $3 \times 100 - 1 = 299$.

Example 8.6

Here is a sequence of numbers.

7, 12, 17, 22, ...

a Work out an expression for the nth term.

b Use your answer to write down the 50th term of the sequence.

Solution

a

Term number (n)	1	2	3	4
Sequence	7	12	17	22

$$+5 \quad +5 \quad +5$$

The common difference is +5 so we look at the values of $5n$.

Term number (n)	1	2	3	4
Sequence	7	12	17	22
$5n$	5	10	15	20

These terms are 2 less than we need so to obtain the nth term we now add 2.

Term number (n)	1	2	3	4
Sequence	7	12	17	22
$5n + 2$	7	12	17	22

The nth term is $5n + 2$.

b The 50th term of the sequence is $5 \times 50 + 2 = 252$.

Practice questions 2

1 Work out an expression for the nth term of each sequence.

  a 3, 8, 13, 18, ...
  b 2, 6, 10, 14, ...
  c 10, 12, 14, 16, ...
  d 5, 9, 13, 17, ...

2 Look at this pattern:
  Line 1: $4^2 - 5 \times 4 + 6 = 2 \times 1$
  Line 2: $5^2 - 5 \times 5 + 6 = 3 \times 2$
  Line 3: $6^2 - 5 \times 6 + 6 = 4 \times 3$
  Line 4: $7^2 - 5 \times 7 + 6 = 5 \times 4$

  a Write down Line 6 of the pattern.
  b Copy and complete Line n for this pattern.
    $(n + 3)^2 - 5 \times \ldots + \ldots = \ldots \times \ldots$

Practice exam questions

1   A sequence begins 1, 3, 7, 15, ...
    The rule for continuing the sequence is shown below.

MULTIPLY THE LAST NUMBER BY 2 AND ADD 1

   a What is the next number in the sequence?
   b This sequence uses the same rule.
      −2, −3, −5, −9, ...
      What is the next number in this sequence?   [AQA (SEG) 1998]

2   Here is a sequence made from a pattern of dots.

   1st pattern   2nd pattern   3rd pattern

   a Copy and complete the table.

Pattern	1	2	3	4	5
Number of dots	5	8	11		

   b How many dots are in the 7th pattern?
   c How many dots are in the nth pattern?
   d Which pattern has 62 dots in it?   [AQA (NEAB) 2002]

3   a What is the next number in this sequence?
         3, 7, 11, 15, ...

   One number in the sequence is x.

   b i   Write, in terms of x, the next number in the sequence.
      ii Write, in terms of x, the number in the sequence before x.   [AQA (SEG) 1999]

4   A sequence begins 1, −2, ...
    The next number in the sequence is found by using the rule:

ADD THE PREVIOUS TWO NUMBERS AND MULTIPLY BY TWO

   Use the rule to find the next *two* numbers in the sequence.   [AQA (SEG) 1999]

5   A man leads a string of donkeys.
    He leads one donkey.
    Altogether they have two heads and six legs.

   He leads two donkeys.
   Altogether they have three heads and ten legs.

   a He leads four donkeys
       i How many heads are there?
       ii How many legs are there?

   b He leads n donkeys.
       i How many heads are there?
       ii How many legs are there?   [AQA (NEAB) 2000]

6 **a** Here are the first three lines of a number pattern.

   Line 1 $1^2 + 2 = 3^2 - 6$

   Line 2 $2^2 + 4 = 4^2 - 8$

   Line 3 $3^2 + 6 = 5^2 - 10$

   Write down the fourth line of this pattern.

   b Here is another pattern.

   Line 1 $1^2 + 2 = 1 \times 3$

   Line 2 $2^2 + 4 = 2 \times 4$

   Line 3 $3^2 + 6 = 3 \times 5$

       i Write down the fourth line of this pattern.
       ii Write down the nth line of this pattern.   [AQA (NEAB) 2000]

7 Here are the first three lines of a number pattern.

   Line 1 $10 \times 1 - 5 = 1 \times 5$

   Line 2 $10 \times 2 - 5 = 3 \times 5$

   Line 3 $10 \times 3 - 5 = 5 \times 5$

   Write down the fourth line of this pattern.   [AQA (NEAB) 2002]

8 Look at this pattern:

   $15^2 - 14^2 = 29$ *row 1*

   $14^2 - 13^2 = 27$ *row 2*

   $13^2 - 12^2 = 25$ *row 3*

   $12^2 - 11^2 = 23$ *row 4*

   a Write down *row 6* of the pattern.
   b Copy and complete this line to give the general rule for this pattern.

   $r^2 - \ldots\ldots\ldots = \ldots\ldots\ldots$ [AQA (NEAB) 2000]

9 Angle facts 1

You will need to know and use the following angle facts.

Vertically opposite angles

Vertically opposite angles are equal.

$x = y$

Angles on a straight line

Angles on a straight line add up to 180°.

$x + y = 180°$

Example 9.1

Find the value of y.

Solution

Angles on a straight line add up to 180°.

Angle $y = 180° - 35°$
$\quad\quad y = 145°$

Angles at a point

Angles at a point add up to 360°.

$p + q + r + s = 360°$
$\quad a + b + c = 360°$

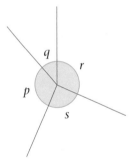

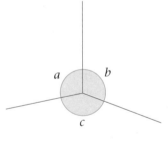

Example 9.2

Find the value of s.

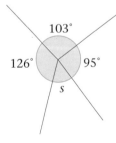

Solution

Angles at a point add up to 360°.
Angle $s = 360° - 126° - 103° - 95°$
$\quad\quad s = 36°$

Practice question 1

1 Work out the missing angles in each part.

a 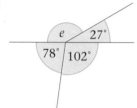 **b** 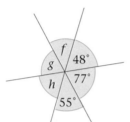 **c**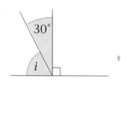

d **e** **f**

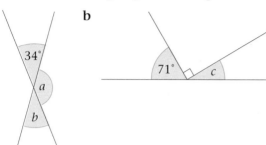

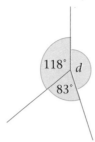

Properties of shapes

You will need to know and use the properties of certain shapes, including the following.

Triangles

A triangle has three interior angles.

● Interior angles of a triangle add up to 180°

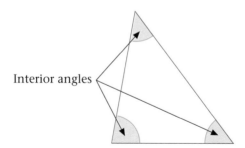

Interior angles

A side of a triangle can be extended.
The angle between a side of a triangle and the side that is extended is called
an **exterior angle**.

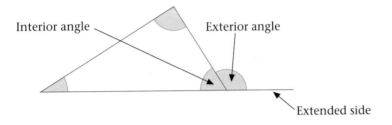

Interior angle · Exterior angle · Extended side

The interior angle and the adjacent exterior angle always sum to 180°.

The exterior angle is equal to the sum of the two interior opposite angles.

$z = x + y$

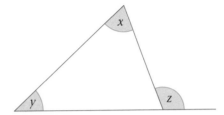

Equilateral triangle

- All sides equal length
- All angles equal to 60°
- Three lines of symmetry
- Rotational symmetry order 3

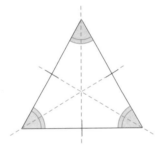

Isosceles triangle

- Two sides equal in length
- Two angles equal
- One line of symmetry
- No rotational symmetry

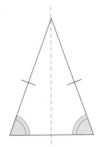

Scalene triangle

- All sides different lengths
- All angles different
- No rotational symmetry

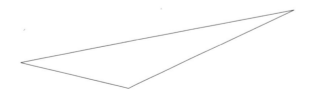

Quadrilaterals

- Any shape with four sides

- Sum of the angles is 360°

You can see that all quadrilaterals have angles that sum to 360° by considering the quadrilateral as two triangles.

Here is a quadrilateral split into two triangles by joining a pair of diagonally opposite corners with a straight line.

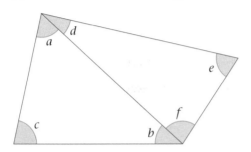

You know that the sum of the angles in a triangle is 180°.
$a + b + c = 180°$
$d + e + f = 180°$

So $a + b + c + d + e + f = 180° + 180°$
$= 360°$

> *Reminder*
> Make sure that you know the properties of a square and a rectangle.

Parallelogram

- Opposite sides parallel

- Opposite sides equal in length

- Opposite angles equal

- Allied angles (x and y) add up to 180°

- Rotational symmetry order 2

Trapezium

- One pair of parallel sides

- Allied angles (x and y) add up to 180°

- No rotational symmetry

> *Reminder*
> Allied angles are angles inside a pair of parallel lines as shown.
>
>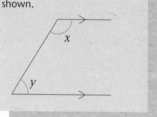

Some trapezia are special because the non-parallel sides are the same length. These are called **isosceles trapezia**.

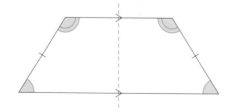

Isosceles trapezium

Example 9.3

Find the value of *x*.

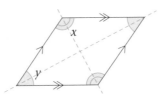

Solution

Allied angles add up to 180°.
Angle *x* = 180° − 46°
　　x = 134°

Rhombus

- Opposite sides parallel
- All sides equal in length
- Opposite angles equal
- Allied angles (*x* and *y*) add up to 180°
- Rotational symmetry order 2

Kite

- Two pairs of adjacent sides equal in length
- One pair of opposite angles are equal
- No rotational symmetry

Polygons

A **polygon** is a shape with many straight sides.

A **regular polygon** has all sides of equal length. All the angles in a regular polygon are equal.

Regular pentagon

Regular hexagon

EXAMINER TIP

It is worth learning the facts that the interior angles of a pentagon add up to 540° and the interior angles of a hexagon add up to 720°.

Exterior and interior angles of regular polygons

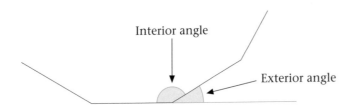

Interior angle

Exterior angle

Reminder
An interior angle is an angle inside the shape and an exterior angle is outside the shape. The interior angle + the exterior angle = 180°.

In a regular polygon the exterior angles are equal, and sum to 360°.

The exterior angle of a regular polygon can be found using the formula:

$$\text{Exterior angle} = \frac{360°}{\text{number of sides}}$$

The interior angle of a regular polygon can be found using the formula:

Interior angle = 180° – exterior angle

If you know the exterior angle of a regular polygon, then you can find the number of sides by rearranging the formula for the exterior angle.

$$\text{Number of sides} = \frac{360°}{\text{exterior angle}}$$

The **sum** of all the interior angles of a regular polygon can be found using the formula:

Sum of the interior angles of a regular polygon = number of sides × interior angle

or using the alternative formula:

Sum of the interior angles of a regular polygon = number of interior angles × interior angle

Example 9.4

The diagram shows a regular polygon with 9 sides (a nonagon).

Calculate:

a the exterior angle (x)

b the interior angle (y)

c the sum of the interior angles.

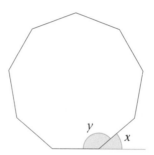

Solution

a To calculate the exterior angle we use the formula:

$$\text{Exterior angle} = \frac{360°}{\text{number of sides}}$$

$$= \frac{360°}{9}$$

$$= 40°$$

b Interior angle + exterior angle = 180°

Interior angle = 180° – exterior angle
            = 180° – 40°
            = 140°

c The sum of the interior angles = interior angle × number of sides
                    = 140° × 9
                    = 1260°

Example 9.5

Complete the table for a shape with 20 sides.

Number of sides	Exterior angle	Interior angle	Sum of interior angles
20			

Solution

Number of sides	Exterior angle	Interior angle	Sum of interior angles
20	$\dfrac{360°}{20} = 18°$	$180° - 18° = 162°$	$162° \times 20 = 3240°$

Practice questions 2

1   For each of these regular shapes calculate:
   i the exterior angle   **ii** the interior angle   **iii** the sum of the interior angles.

   a pentagon (5 sides)   **b** hexagon (6 sides)   **c** octagon (8 sides)   **d** decagon (10 sides)

2   Copy and complete the table for regular polygons.

Number of sides	Exterior angle	Interior angle	Sum of interior angles
7			
9			
11			
12			

Drawing regular polygons

A regular polygon can be drawn using the following method.

Step 1

Draw a circle.

Step 2

Divide 360° by the number of sides of your regular polygon.

For example, for a hexagon divide 360° by 6 (= 60°).

Step 3

Divide the circle into equal sectors where the angle at the centre of the circle for each sector is equal to the angle calculated in step 2.

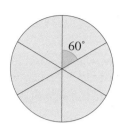

For example, for a hexagon, divide the circle into 6 equal sectors with angles of $360° \div 6 = 60°$ at the centre of the circle.

Step 4

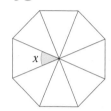

Join up the points where the sector lines meet the circumference of the circle to form your polygon.

The angle at the centre of a regular polygon

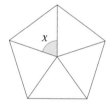 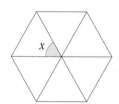 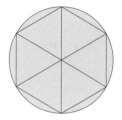

The angle x is called the angle at the centre. In a regular polygon all the angles at the centre are equal.
The angle at the centre can be calcualted using this formula:

$x = \dfrac{360°}{n}$, where n is the number of sides.

Example 9.6

Find the angle at the centre of the octagon.

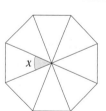

Solution

Using the formula: $x = \dfrac{360}{n}$

An octagon has 8 sides so $n = 8$.

$x = \dfrac{360°}{8}$

$x = 45°$

Practice question 3

1   Use the circle method to form the following regular shapes.

   a pentagon   **b** octagon   **c** decagon

Practice exam questions

1   Triangle *ABC* is isosceles.
AB = AC.
Work out the size of angle *x*.

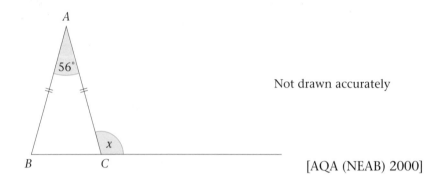

Not drawn accurately

[AQA (NEAB) 2000]

2   Triangle *PQR* is isosceles.
PQ = PR.
Calculate the size of the angle
marked *x*.

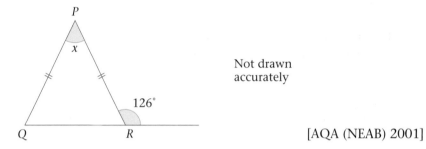

Not drawn
accurately

[AQA (NEAB) 2001]

3   **a**   The diagram shows the rectangle *ABCD*.
M is the mid point of *DC*.
Angle *AMB* = 80°.
AM = MB.

Work out the sizes of angles *x* and *y*.

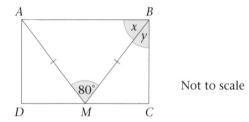

Not to scale

   b   The diagram shows a quadrilateral *PQRS*.
PQ = QR and *PS = SR*.

   i   Which of the following correctly describes
the quadrilateral *PQRS*?

   Diamond   Kite   Rhombus
   Parallelogram   Trapezium

   ii   Angle *PSR* = 42° and angle *QRS* = 100°.
Work out the sizes of angles *p* and *q*.

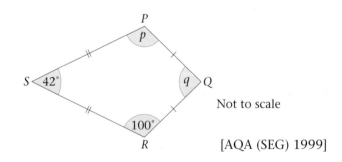

Not to scale

[AQA (SEG) 1999]

4 The diagram shows a kite.

Calculate the size of the angle marked *z*.

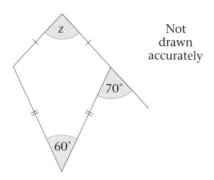

Not drawn accurately

[AQA (NEAB) 2001]

5 The diagram shows a slab in the shape of a pentagon.
It has one line of symmetry.

The slabs are used to make a path.
All the slabs are the same shape and size.

The slabs are arranged as shown in the diagram.

Two angles are given on the diagram.
Work out the size of the angle marked *c*.

Not drawn accurately

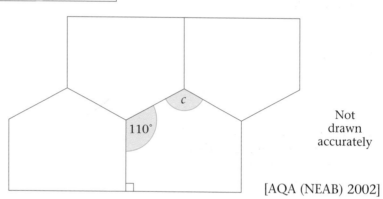

Not drawn accurately

[AQA (NEAB) 2002]

6 The diagram shows a regular octagon with centre *O*.

a Work out the size of angle *x*.
b Work out the size of angle *y*.

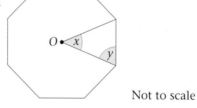

Not to scale

[AQA (SEG) 1998]

7 The diagram shows a shape which consists of two regular polygons.
Work out the size of angle *x*.

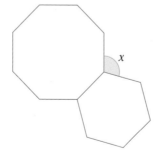

Not to scale

[AQA (SEG) 1999]

10 Constructing triangles

You may be asked to draw a triangle using a ruler and protractor. There are two different cases that can arise.

Case 1: Given two sides and the included angle

Example 10.1

Draw a triangle ABC with $AB = 7$ cm, $AC = 5$ cm and angle $CAB = 65°$.

Solution

Often, drawing a rough sketch of the triangle before you start will help you to see what to do in the construction.

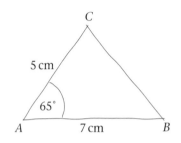

Step 1 Draw the side AB 7 cm long.

Step 2 Measure the angle CAB as 65°.

Step 3 Faintly draw in the side AC.

Step 4 Measure AC as 5 cm and mark C.

Step 5 Join C to B to complete the triangle.

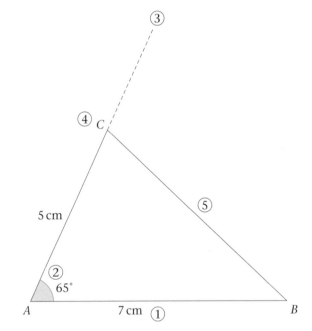

Case 2: Given one side and two angles

Example 10.2

Draw a triangle *PQR* with *PQ* = 9 cm, angle *QPR* = 55° and angle
PQR = 40°.

Solution

Before you start, draw a
rough sketch to see what to
do.

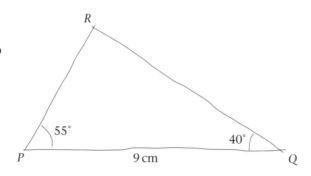

Step 1 Draw side *PQ* 9 cm long.

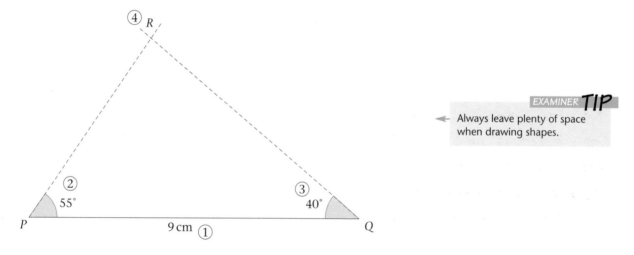

EXAMINER *TIP*
Always leave plenty of space
when drawing shapes.

Step 2 Measure and mark angle *QPR* as 55°.
Draw in a faint construction line.

Step 3 Measure and mark angle *RQP* as 40°.
Draw in a faint construction line.

Step 4 Where the two construction lines cross
label that point *R*.

Step 5 Change the construction lines into bold
lines to complete the triangle *PQR*.

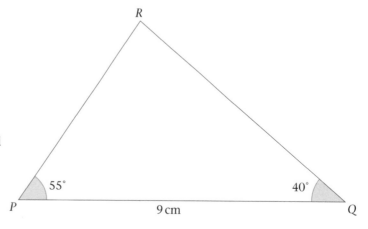

Practice question

1   Make accurate drawings of the following triangles.

    a △*ABC*, *AB* = 10 cm, ∠*BAC* = 38°, *AC* = 8 cm
    b △*PQR*, *PR* = 13 cm, ∠*QRP* = 65°, *RQ* = 7 cm
    c △*XYZ*, *XY* = 15 cm, ∠*ZXY* = 75°, ∠*XYZ* = 35°
    d △*RST*, *ST* = 10 cm, ∠*RST* = 55°, ∠*RTS* = 65°

11 Circles

You are expected to already know the words associated with circles.

The **centre** of the circle is the same distance from every point of the circumference (the circle's edge).

The **diameter** is the distance of a straight line going from any point on the circle, through the centre, to another point on the circle.

 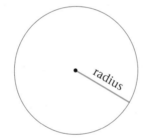

The **radius** is the distance of a straight line going from any point on the circle to the centre. The radius is half the length of the diameter.

A **chord** is a straight line that joins any two points on the circle.

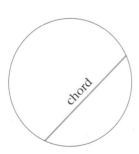 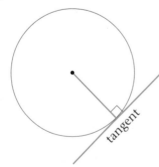

A **tangent** is a straight line outside the circle that only touches the edge of the circle at one point. If you joined that point to the centre of the circle the angle between the tangent and the second line would be 90°.

An **arc** is a section of the curve of the edge of a circle.

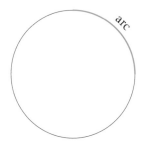  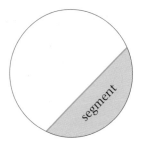

A **sector** is a part of the whole circle made by joining two points on the circle to the centre.
A **segment** is the area created by a chord.

Circumference of a circle

The **circumference** is a special name for the perimeter.

To calculate the circumference of a circle use these formulae:

Circumference = 2 × π × radius or Circumference = π × diameter
These are often shortened to:
$C = 2\pi r$ or $C = \pi d$

The exam paper tells you to use π = 3.14 or the π button on your calculator, unless otherwise instructed in the question.

Sometimes, usually on the non-calculator paper, you will be asked to leave your answer in terms of π. In Example 11.1 the answer in terms of π is 16π cm. This is the exact answer. If you are not told how to write your answer it is sensible to round it off to 2 significant figures.

Example 11.1

a Calculate the circumference of a circle of radius 8 cm, leaving your answer in terms of π.

b Calculate the circumference of a circle of diameter 10 cm.

Solution

a Using the formula:
$C = 2\pi r$
  $= 2 \times \pi \times 8$
  $= 16\pi$ cm
So the circumference is 16π cm.

b Using the formula:
$C = \pi d$
  $= \pi \times 10$
  $= 31.41592...$
  $= 31$ cm (2 s.f.)

EXAMINER TIP
Make sure you know how to enter π on your calculator.

Reminder
See Chapter 1 for help with significant figures.

Finding the radius given the circumference

You may be given the circumference of a circle and then be asked to find either the radius or the diameter.

Use the same formula and then rearrange to make the radius (or diameter) the subject.

Example 11.2

Calculate the diameter of a circle with a circumference of 20 metres.

Solution

Using $C = 2\pi r$	Using $C = \pi d$
$20 = 2 \times \pi \times r$	$20 = \pi \times d$
Dividing both sides by 2π gives:	Dividing both sides by π gives:
$\dfrac{20}{2\pi} = r$	$\dfrac{20}{\pi} = d$
3.18... m = radius	6.37 m = diameter
To obtain the diameter: diameter = 2 × radius = 2 × 3.18... = 6.37 m	

EXAMINER **TIP**
Do not round your answer until the end of the calculation.

Example 11.3

A wheel of radius 30 cm makes 10 complete revolutions.
How far does it travel?

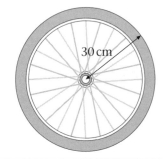
30 cm

Solution

Using $C = 2\pi r$	Using $C = \pi d$
For 1 revolution: Distance travelled = 1 full circumference Circumference = 2 × π × 30 = 60π cm or 188 cm	For 1 revolution: Distance travelled = 1 full circumference Circumference = π × 60 = 60π cm or 188 cm
For 10 revolutions: Distance travelled = 10 × 60π or 10 × 188 = 600π cm or 1880 cm = 18.8 m	For 10 revolutions: Distance travelled = 10 × 60π or 10 × 188 = 600π cm or 1880 cm = 18.8 m

Practice questions 1

1 Calculate the circumference of each of the following circles, leaving your answer in terms of π.

    a radius = 4 cm   **b** radius = 17 cm
    c diameter = 2 m   **d** diameter = 4.8 cm

2 Calculate the radius of each of the following circles, giving your answers to 2 significant figures.

    a circumference = 100 cm   **b** circumference = 28 cm
    c circumference = 18.9 cm   **d** circumference = 450 cm

3 Calculate the diameter of each of the following circles, giving your answers to 2 significant figures.

    a circumference = 43 cm   **b** circumference = 120 cm
    c circumference = 56.7 cm   **d** circumference = 355 cm

Area of a circle

> *Reminder*
> The units for area are mm², cm², m², km², etc.

You will be expected to find the area of a circle using this formula:

Area = $\pi \times$ radius^2 or $A = \pi r^2$

> EXAMINER **TIP**
> When calculating r^2 remember that it means $r \times r$, e.g. $4^2 = 4 \times 4 = 16$.

Example 11.4

Calculate the area of a circle of radius 10 cm.

Solution

> *Reminder*
> Use $\pi = 3.14$ or the π button on your calculator.

Using the formula: $A = \pi r^2$
                      $= \pi \times 10^2$
                      $= 100\pi$ cm^2 or 314 cm^2

Example 11.5

Calculate the area of a circle of diameter 7 cm.

Solution

The diameter is 7 cm. So the radius is $7 \div 2 = 3.5$ cm.

Using the formula: $A = \pi r^2$
                      $= \pi \times 3.5^2$
                      $= 12.25\pi$ cm^2 or 38.5 cm^2 (to 1 decimal place)

Example 11.6

A shape is made up of a square of side 6 cm and two semicircles of diameter 6 cm, as shown. Calculate the area of the shape.

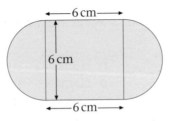

Solution

The area of the square $= 6 \times 6$
$ = 36 \text{ cm}^2$
The two semicircles can be combined to make a complete circle of radius 3 cm.
So the area of the two semicircles $= \pi \times 3^2$
$ = 28.3 \text{ cm}^2$
So the total area of the shape $= 36 + 28.3$
$ = 64.3 \text{ cm}^2$

To find the radius or diameter of a circle given the area

To find the radius or diameter use the area formula but then work backwards.

Reminder

$A = \pi r^2$

Example 11.7

A circle has an area of 50 cm^2. Calculate the radius.

Reminder

If you are asked to find the diameter remember it is $2 \times$ radius.

Solution

Step 1 Using the formula: $A = \pi r^2$
$ 50 = \pi \times r^2$

Step 2 Dividing both sides by π:
$$\frac{50}{\pi} = r^2$$

Step 3 Taking the square root gives:
$$r = \sqrt{\frac{50}{\pi}} = 3.99 \text{ cm (or 4 cm to the nearest cm)}$$

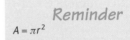

EXAMINER **TIP**

Check the answer line on the exam paper to see if the units are given. If they are not given there will be a mark awarded for stating them.

Practice questions 2

1 Calculate the area of each of the following circles

    i   leaving your answer in terms of π
    ii giving your answer to 2 significant figures.
        a   radius $= 5$ cm   **b**   radius $= 23$ cm
        c   diameter $= 8$ m   **d**   diameter $= 17.2$ cm

2 Calculate the radius of each of the following circles, giving your answers to 2 significant figures.

        a   area $= 120$ cm^2 **b**   area $= 60$ cm^2
        c   area $= 82.3$ cm^2 **d**   area $= 4500$ cm^2

Practice exam questions

1   **a** Calculate the circumference of a circle of diameter 26 cm.
    State the units of your answer.
  b Calculate the area of a circle of radius 2.5 cm.
    State the units of your answer.

[AQA (NEAB) 2001]

2   **a** Circular labels of diameter 10 cm are stuck on a sheet.
    The sheet measures 1.4 metres by 0.8 metres.

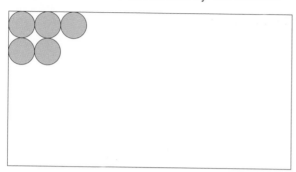

Not drawn
accurately

How many labels will fit on the sheet?

  b A circle has diameter 10 cm.

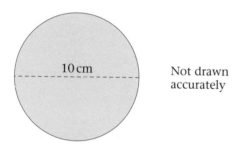

10 cm   Not drawn
            accurately

Calculate the circumference of the circle.
Take the value of π as 3.14.

[AQA (NEAB) 2002]

12 Plans and elevations

Plan and elevation

Here is a picture of a garden shed.

You can see the roof, the front with a window and the side with a door. These can be drawn as three separate two-dimensional views:

- the view from **above**, called the **plan view**, showing the roof
- the view from the **front**, called the **front elevation**
- the view from the **side**, called the **side elevation**.

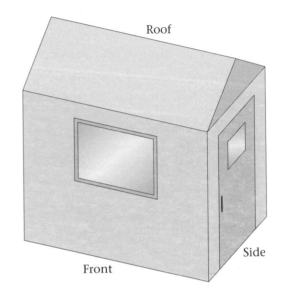

Roof

Front

Side

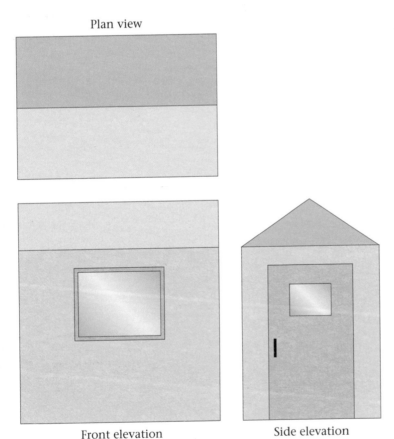

Plan view

Front elevation

Side elevation

Shapes made of cubes

Often in questions you will be given a shape made of cubes and asked to draw the plan and elevations of it. Here is a three-dimensional shape.

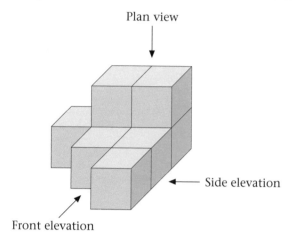

Plan view

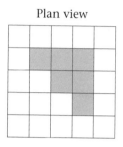
Plan view

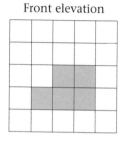
Front elevation   Side elevation

Side elevation

Front elevation

We can again draw two-dimensional diagrams showing the three views.

Example 12.1

A solid is made from a cube of side 2 cm and three cubes of side 1 cm as shown.

a Draw the plan view, front elevation and side elevation of the solid on square grids.

b Work out the volume of the solid.

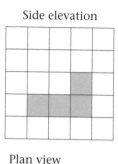
Plan view

Side elevation

Front elevation

Solution

a

Plan view

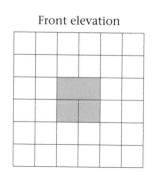
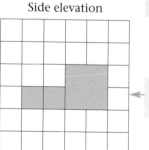
Front elevation   Side elevation

b Volume of large cube
= 2 × 2 × 2 = 8 cm³

Volume of each small cube
= 1 × 1 × 1 = 1 cm³

So, total volume
= 8 + 1 + 1 + 1 = 11 cm³

EXAMINER **TIP**
If the units for volume are not given, remember to write them with your answer as it will be worth a mark.

EXAMINER **TIP**
You will usually be given a grid to draw your answer on.

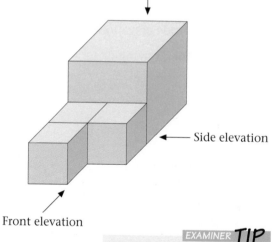

Drawing shapes from plans and elevations

You may be asked to draw 3D objects from 2D plans and elevations.

Example 12.2

The plan view, front elevation and side elevation of a 3D solid made up of cubes is shown below.

Draw the solid.

Plan view

Front elevation   Side elevation

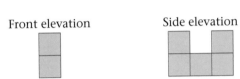

Solution

The plan view shows how the cubes are arranged on the base.

The front and side elevations show that the highest cubes are two blocks high.

Completing the drawing gives:

Practice questions

1 Here are some drawings of solid objects together with the plan view of each. Match the object with its plan view.

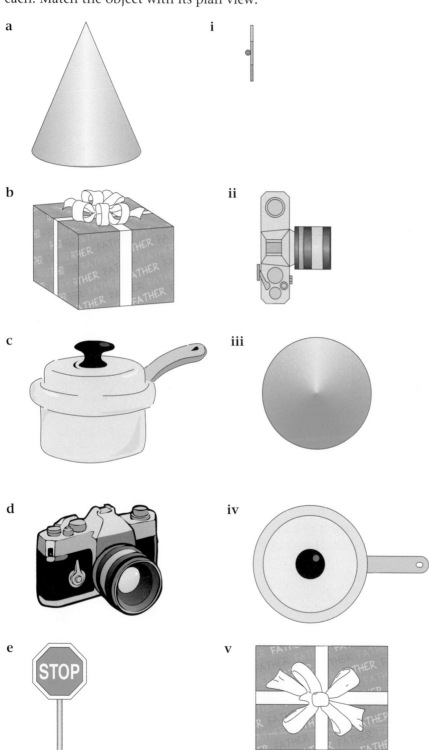

a

i

b

ii

c

iii

d

iv

e

v

2   For each solid, draw on centimetre grid
    paper:

    ● the plan view
    ● the front elevation
    ● the side elevation.

a    **b**

c 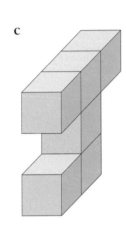   **d** 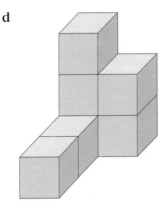   **e**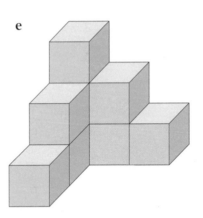

f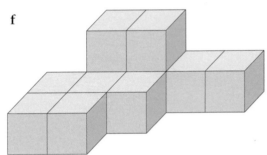

3   The plan view, front elevation and side elevation of a 3D solid made up of
    cubes is shown below. On isometric paper draw the solid.

Plan view

Front elevation

Side elevation

13 Solving linear equations

An **equation** is a mathematical statement containing an equals sign.

In a **linear equation** letters are only to the power 1. This means they would not contain, for example, x^2 but they could contain x.

Some examples are:

$$x + 1 = 7$$
$$2y - 3 = 11$$
$$5(a - 6) = 10$$
$$\frac{p + 1}{2} = 8$$
$$2q + 3 = q - 5$$
$$\frac{4x - 1}{2} = 6x$$

Solving an equation means finding the value of the letter which satisfies the equation.

You can do this by carrying out operations $(+, -, \times, \div)$ to **both sides** of the equation in order to have the letter on one side and the numbers on the other side of the equation.

Example 13.1

Solve the equation $x + 1 = 7$.

Solution

$$x + 1 = 7$$

Subtract 1 from both sides $\quad x + 1 - 1 = 7 - 1$

$$x = 6$$

Example 13.2

Solve the equation $2y - 3 = 11$.

Solution

$$2y - 3 = 11$$

Add 3 to both sides $\quad 2y - 3 + 3 = 11 + 3$

$$2y = 14$$

Divide both sides by 2 $\quad y = 7$

Example 13.3

Solve the equation $5(a - 6) = 10$.

Solution

Method A (expanding the brackets)

$$5(a - 6) = 10$$

Removing brackets $\qquad 5a - 30 = 10$

Add 30 to both sides $\qquad 5a = 40$

Divide both sides by 5 $\qquad a = 8$

Method B (dividing first)

$$5(a - 6) = 10$$

Divide both sides by 5 $\qquad a - 6 = 2$

Add 6 to both sides $\qquad a = 8$

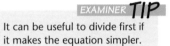 EXAMINER **TIP**
It can be useful to divide first if it makes the equation simpler.

Example 13.4

Solve the equation $\dfrac{p + 1}{2} = 8$.

Solution

$$\frac{p + 1}{2} = 8$$

Multiply both sides by 2 $\qquad p + 1 = 16$

Subtract 1 from both sides $\qquad p = 15$

Example 13.5

Solve the equation $2q + 3 = q - 5$.

Solution

$$2q + 3 = q - 5$$

Subtract q from both sides $\qquad q + 3 = -5$

Subtract 3 from both sides $\qquad q = -8$

Example 13.6

Solve the equation $\dfrac{4x - 1}{2} = 6x$.

Solution

$$\frac{4x - 1}{2} = 6x$$

Multiply both sides by 2 $\qquad 4x - 1 = 12x$

Subtract $4x$ from both sides $\qquad -1 = 8x$

Divide both sides by 8 $\qquad -\dfrac{1}{8} = x$

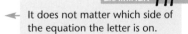 EXAMINER **TIP**
It does not matter which side of the equation the letter is on.

Practice question

1   Solve each equation.

   a $5r + 1 = 16$ **b** $8s + 7 = 11$ **c** $\dfrac{t}{4} = 12$

   d $2x + 4 = 10$ **e** $3y - 7 = 11$ **f** $\dfrac{z + 1}{3} = 5$

   g $3a + 6 = a - 7$ **h** $5(2b + 3) = 35$ **i** $12 - m = 2m + 21$

   j $3(x + 5) = 4x - 1$ **k** $12 - h = 3 - 4h$

Practice exam questions

1   **a** Solve the equation $2x + 5 = 8$.

   b Solve the equation $10x + 4 = 2x + 8$.   [AQA (NEAB) 2002]

2   Solve the equation $6x + 3 = 5 - 2x$.   [AQA (NEAB) 2002]

3   Solve $5x - 3 = 3x + 7$.   [AQA (SEG) 1999]

4   **a** Solve the equation $7x - 2 = 4x - 1$.

   b Solve the equation $\dfrac{x - 4}{6} = 3$.   [AQA (NEAB) 2001]

5   Solve the equation $\frac{1}{3}x = 2$.   [AQA (SEG) 1999]

6   Solve these equations.

   a $5x - 2 = 13$

   b $3(2x - 1) = 9$ [AQA (SEG) 2000]

7   Solve the equations.

   a $7x - 13 = 5(x - 3)$

   b $\dfrac{10 - x}{4} = x - 1$ [AQA (NEAB) 2000]

14 Forming and solving linear equations

Inverse operations

There are four common operations: +, −, ×, ÷.

It is useful to know the words to explain these operations.

+	−	×	÷
Add Sum Total Altogether	Subtract Take away Difference	Multiply Times Product	Divide Share by

Each of these operations has an inverse (or opposite) operation, e.g. the inverse of adding is subtracting, the opposite of multiplying is dividing.

Operation	Inverse (opposite) operation
+	−
−	+
×	÷
÷	×

There are many other operations that have an inverse operation, e.g.

Operation	Inverse (opposite) operation
Square root $\sqrt{}$	Square
Square	Square root $\sqrt{}$
Cube root $\sqrt[3]{}$	Cube
Cube	Cube root $\sqrt[3]{}$

Example 14.1

In 3 years time John will be 19 years old. How old is John now?

Solution

You can often use algebra to help solve a problem.

Let John's age be x years old now.

You want to find x.

So set up this equation $x + 3 = 19$

3 has been added to x to give a total of 19.

The opposite of addition is subtraction.

Subtract 3 from both sides of the equation.

$$x + 3 - 3 = 19 - 3$$

As $3 - 3 = 0$, this can be removed from the left-hand side of the equation leaving you with x alone.

So $x = 19 - 3$

$$x = 16$$

John is 16 years old now.

> **Reminder**
> An equation is like a balance; you have to do exactly the same operation to both sides of the equation to keep it balanced.

Example 14.2

Wendy receives a cheque for £18.56. She calculates that after depositing the cheque in her bank account she will have a total of £74.78. How much does Wendy have in her bank account before she deposits the cheque?

Solution

Using algebra to help solve this problem, let £x be the amount in Wendy's bank account before she deposits the cheque.

Set up an equation   £x + £18.56 = £74.78

(Since all of the equation is in the same units you can ignore the £ sign until you get to the answer to the question.)

So $x + 18.56 = 74.78$

The 18.56 is added to the x to give a total of 74.78.

The opposite of addition is subtraction so subtract 18.56 from both sides of the equation.

$$x + 18.56 - 18.56 = 74.78 - 18.56$$

As $18.56 - 18.56 = 0$, this can be removed from the left-hand side of the equation leaving you with x alone.

This gives $x = 74.78 - 18.56$

$$x = 56.22$$

Wendy has £56.22 in her account before she deposits the cheque.

Flow diagrams

Operations and inverse operations can also be illustrated by using flow diagrams.

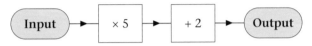

This represents starting with a number, multiplying by 5 and then adding 2.

Example 14.3

Look at the flow diagram above.

a What is the output when the input is 3?

b What is the input when the output is –5?

Solution

a Starting with 3 and multiplying by 5 gives 15. Adding 2 gives a final output of 17.

b Working backwards the flow chart becomes:

So if the output is –5, subtracting 2 gives –7. Now dividing by 5 gives $\frac{-7}{5} = -1.4$.

Practice questions 1

1 Find x from each of the following equations.

    a $x + 12 = 25$ b $x + 19 = 36$

    c $x + 24 = 76$ d $x + 33 = 125$

    e $43 = x + 12$ f $55 = x + 28$

    g $134 = x + 76$

2 Jenny asked Peter to think of a number and add 12 to it. Peter says his answer is 44. What number did Peter think of?

Solving more complex problems using equations

To **form an equation** is to interpret a statement or diagram, sometimes using formulae from other parts of the syllabus. You will usually then be asked to solve the equation.

You may need to recall the formulae for perimeter, area and sums of interior angles of polygons.

Example 14.4

a Write an expression for the perimeter of this triangle.
b The perimeter of the triangle is 30 centimetres. Work out the value of x and state the length of the longest side.

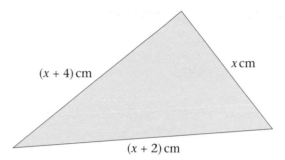

$(x + 4)$ cm

x cm

$(x + 2)$ cm

Solution

a The perimeter is the total length of the sides of the triangle.
$$P = x + (x + 2) + (x + 4)$$
$$= 3x + 6$$

b $$3x + 6 = 30$$
Subtract 6 from both sides $\quad 3x = 24$
Divide both sides by 3 $\qquad\quad x = 8$

Longest side is $x + 4 = 8 + 4$
$$= 12 \text{ cm}$$

Example 14.5

The angles of a triangle are x, $2x$ and $3x$.

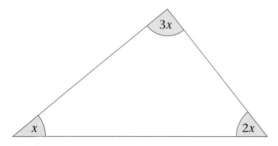

$3x$

x

$2x$

Work out the size of each angle.

Solution

The values of the angles in a triangle add up to $180°$.

$$x + 2x + 3x = 180°$$
$$6x = 180°$$
Dividing both sides by 6 $\qquad x = 30°$

The angles are

$x = 30°$
$2x = 2 \times 30° = 60°$
$3x = 3 \times 30° = 90°$

Check your working to see if all the angles add to $180°$: $30° + 60° + 90° = 180°$.

Example 14.6

I think of a number, double it and subtract 6. The answer is 18.
What number am I thinking of?
Show your working.

Solution

Let the number be x.
Setting up the equation gives

$$2x - 6 = 18$$

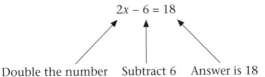

Double the number   Subtract 6   Answer is 18

Add 6 to both sides	$2x = 24$
Divide both sides by 2	$x = 12$

Example 14.7

The cost of hiring a car for one day is £40 plus 20p per mile.

a Calculate the total cost of hiring the car for one day and travelling 120 miles.

b Write an expression for the total cost of hiring the car for one day and travelling m miles.

Solution

a The cost of hiring the car is £40.
The cost of travelling 120 miles is:
$120 \times 20p = £24$.
The total cost is: $£40 + £24 = £64$.

b The cost of hiring the car is £40.
The cost of travelling m miles is:
$m \times 20p = £0.20m$.
The total cost is $£40 + £0.20m$.

This may be written as $£\left(40 + \dfrac{m}{5}\right)$.

> **Reminder**
>
> $0.20m$ is the same as $\dfrac{m}{5}$.

Practice questions 2

1 The length of a rectangle is double its width.
The area of the rectangle is 98 cm^2.
Work out the length of the rectangle.

2 A pentagon has angles x, $3x$, $30°$, $x + 70°$, and $x + 20°$.

    a Write down an expression
       for the sum of the angles
       in terms of x.

    b Write down an equation
       in terms of x.

    c Find the value of x.

    d Work out the size of the
       largest angle.

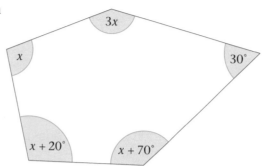

> **Reminder**
> The angles in a pentagon add up to $540°$.

Practice exam questions

1 Here is a flow diagram.

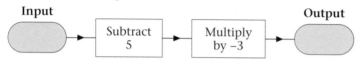

    a What is the output when the input is 3?

    b What is the input when the output is -21?

                                            [AQA (NEAB) 2001]

2 **a** This is a number machine. You start with 36.
       What goes in the answer box?

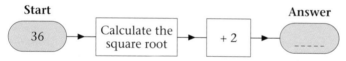

    b This is a different number machine. You start with 27.

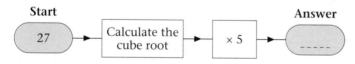

       What goes in the answer box?

    c This is an algebra machine. You start with x.
       Write down the algebraic expression which goes in the answer box.

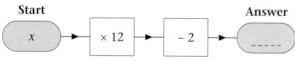

    d This is a different algebra machine.
       The answer box has been completed.
       Copy and complete the diagram.

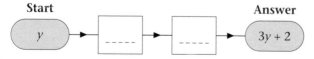

                                            [AQA (NEAB) 2000]

3 A formula for converting degrees Celsius to degrees Fahrenheit is shown by the flow chart. Convert −5° Celsius to degrees Fahrenheit.

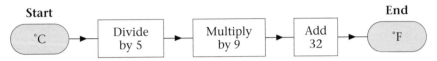

[AQA (NEAB) 2002]

4 Adrian has three regular polygons: *A*, *B* and *C*.
A has *x* sides. *B* has $(2x - 1)$ sides. *C* has $(2x + 2)$ sides.

   a Write an expression, in terms of *x*, for the total number of sides of these three polygons. Write your answer in its simplest form.

   The three polygons have a total of 16 sides.

   b i Form an equation and hence find the value of *x*.
   ii Use your value of *x* to find the number of sides of polygon *B*.

[AQA (SEG) 1999]

5 The area of the parallelogram is equal to the area of the trapezium. Find the value of *x*.

Diagrams not drawn accurately

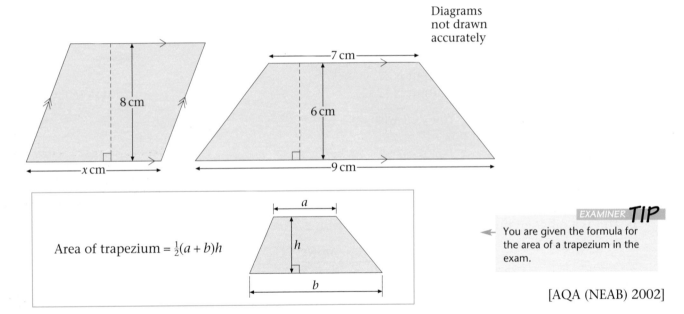

Area of trapezium = $\frac{1}{2}(a + b)h$

EXAMINER **TIP**

◄ You are given the formula for the area of a trapezium in the exam.

[AQA (NEAB) 2002]

6 A small paving slab weighs *x* kilograms.
A large paving slab weighs $(2x + 3)$ kilograms.

   a Write an expression, in terms of *x*, for the total weight of 16 small slabs and 4 large slabs.
   Give your answer in its simplest form.

   The total weight of the slabs is 132 kilograms.

   b Write down an equation and find the value of *x*.

[AQA (SEG) 1999]

15 Linear graphs

Plotting straight line graphs from a table of values

You will be required to plot points on graph paper accurately and join them to form the straight line. Questions may ask you to complete a **table of values** or you may be expected to calculate points of your choice in order to draw the graph. You can use the **cover-up method** to find corresponding values of x and y for a particular equation, e.g. $x + y = 7$, x could equal 0. You can cover it up, so $y = 7$ when $x = 0$.

Example 15.1

a Complete the table of values for $y = 2x - 3$.

x	−2	−1	0	1	2
y		−5			1

b Draw the graph of $y = 2x - 3$ for values of x from −2 to 2.

Solution

a When $x = -2$, $y = 2 \times -2 - 3$
$\qquad\qquad\qquad\quad = -7$
  When $x = 0$, $y = 2 \times 0 - 3$
$\qquad\qquad\qquad = -3$
  When $x = 1$, $y = 2 \times 1 - 3$
$\qquad\qquad\qquad = -1$

x	−2	−1	0	1	2
y	−7	−5	−3	−1	1

> *Reminder*
> The x-value tells you how far across the point is, the y-value tells you how far up the point is.

b
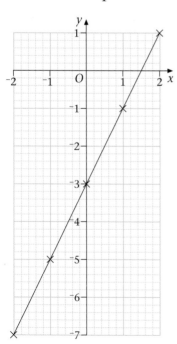

> EXAMINER **TIP**
> Two or three points is enough to be able to draw the straight line so you do not have to work out lots of points unless you are asked to complete a table.

> EXAMINER **TIP**
> Take care when plotting points as they have to be plotted accurately to avoid losing marks. Always use a ruler when drawing straight lines.

Practice questions 1

1   Draw the graph of $y = 2x + 1$ for values of x from −1 to 3.

2   Draw the graph of $y = -x + 2$ for values of x from −2 to 2.

3   Draw the graph of $y = \frac{1}{2}x - 1$ for values of x from 0 to 6.

Gradient of a straight line

The **gradient** of a line measures the steepness of the line on a graph.
Lines can have a **positive** gradient, a **negative** gradient or **zero** gradient.

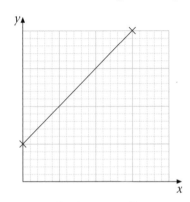 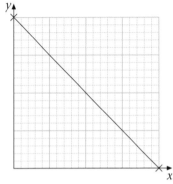 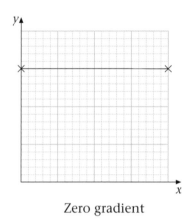

Positive gradient   Negative gradient   Zero gradient

Finding the gradient of line

The gradient of a line can be found using the formula:

$$\text{Gradient} = \frac{\text{Change in } y\text{-values between two points}}{\text{Change in } x\text{-values between two points}}$$

Example 15.2

A line passes through the points (0, 1) and (5, 3) as shown.

Work out the gradient of the line.

Solution

$$\begin{aligned}
\text{Gradient} &= \frac{\text{Change in } y\text{-values}}{\text{Change in } x\text{-values}} \\[2mm]
&= \frac{3 - 1}{5 - 0} \\[2mm]
&= \frac{2}{5}
\end{aligned}$$

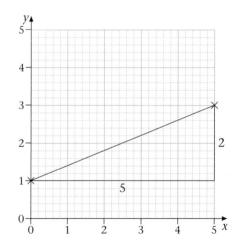

Linear equations

A **linear equation** will be represented by a straight line when drawn on graph paper. You need to be able to recognise and draw straight line graphs which are parallel to the *x*-axis or the *y*-axis.

The line shown on the graph has the equation $x = 1$, because all the points on the line have an *x*-coordinate equal to 1.

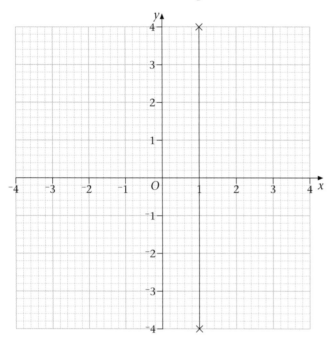

The line shown on this graph is the line $y = 2$, because all the points on the line have a *y*-coordinate equal to 2.

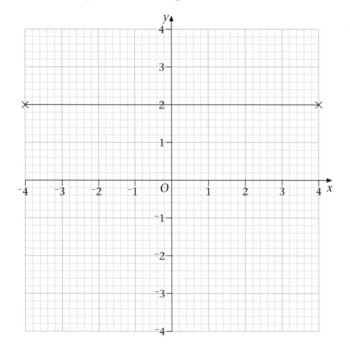

Linear equations that are not parallel to an axis will have a term in y, a term in x and often a constant term.

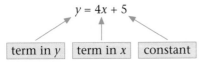

$y = 4x + 5$

| term in y | term in x | constant |

$2x + 3y + 4 = 0$

| term in x | term in y | constant |

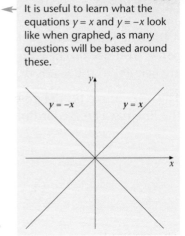

EXAMINER *TIP*

It is useful to learn what the equations $y = x$ and $y = -x$ look like when graphed, as many questions will be based around these.

Linear equations written in the form $y = mx + c$

When a linear equation is written in the form $y = mx + c$, e.g. $y = 3x + 2$, m is the value of the gradient and c is the y-intercept (the point where the line cuts the y-axis).

Example 15.3

Write down the gradient m and the y-intercept c of the line $y = 3x + 2$.

Solution

Compare $y = 3x + 2$ with $y = mx + c$.

The gradient $m = 3$ and the y-intercept $c = +2$.

Example 15.4

Write down the gradient m and the y-intercept c of the line $2x - 3y = 4$.

Solution

The equation has to be rearranged into the form $y = mx + c$.

$$2x - 3y = 4$$

Adding $3y$ to both sides gives $\qquad 2x = 3y + 4$

Subtracting 4 from both sides gives $\qquad 2x - 4 = 3y$

Dividing each term by 3 gives $\qquad \dfrac{2}{3}x - \dfrac{4}{3} = y$

This can be rewritten as $y = \dfrac{2}{3}x - \dfrac{4}{3}$, which is now in the correct form

$y = mx + c$.

The gradient $m = \dfrac{2}{3}$ and the y-intercept $c = -\dfrac{4}{3}$.

Sketching graphs and finding an equation given two points

You may be asked to sketch a graph given the equation or you may be asked to work out the equation of a line given two points.

Example 15.5

The graph shows a straight line passing through the points $A(0, 4)$ and $B(5, 7)$.

Work out the equation of the line.

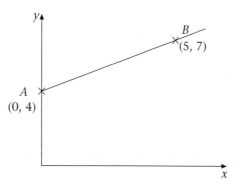

Solution

The gradient
$m = \dfrac{\text{Change in } y\text{-values}}{\text{Change in } x\text{-values}} = \dfrac{3}{5}$ and
the y-intercept $c = +4$.

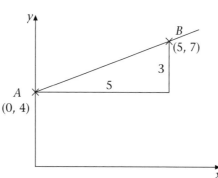

The equation of the line is

$y = \dfrac{3}{5}x + 4.$

EXAMINER *TIP*
Full marks will be given in the exam for leaving the equation in this form.

Example 15.6

The graph shows a straight line passing through the points $A(0, 6)$ and $B(5, 4)$.

Work out the equation of the line.

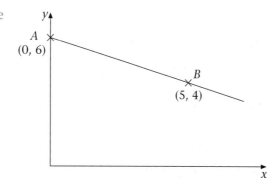

Solution

The gradient $m = -\dfrac{2}{5}$ and the
y-intercept $c = +6$.

The equation of the line is

$y = -\dfrac{2}{5}x + 6.$

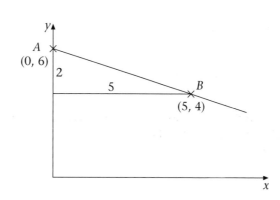

Reminder
When lines slope this way the gradient is negative.

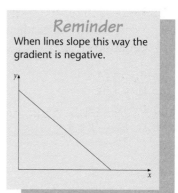

Example 15.7

Sketch the graph of $y = 2x + 3$.

Solution

Look at the equation. Since it is in the form $y = mx + c$, you can tell immediately what the gradient and the y-intercept are. The gradient, m, is 2. The y-intercept, c, is 3.

Use this information to draw a line on a graph with unlabelled axes. The line should look steeper than if it was a $y = x$ line, as the gradient is steeper. The line should also cross the y-axis on the positive side. Label the point where it crosses the y-axis as 3.

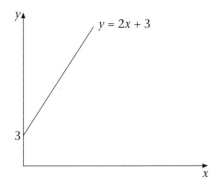

Practice questions 2

1   Sketch graphs of the following lines.

    a $x = 4$
    b $x = -2$
    c $y = 3$
    d $y = -5$

2   Write down the gradient m and the y-intercept c of the following equations.

    a $y = 2x + 1$
    b $y = 5x$
    c $4y - x = 5$
    d $3x + 7y = 4$
    e $2x - y = 3$

3 Work out the equation of the line for each of the following graphs.

a
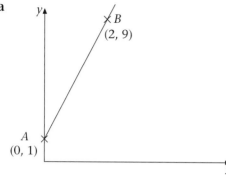

b
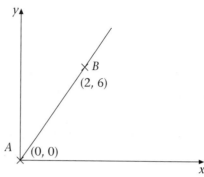

c
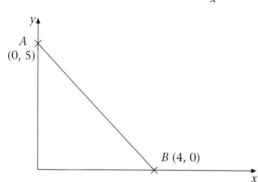

d
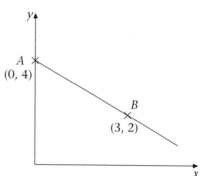

Parallel lines

Two lines that are **parallel** have the same gradient. By writing two equations in the form $y = mx + c$, and comparing the values of m of each line it is possible to state whether the two lines are parallel or not parallel.

The equations of two straight lines are $4y - 8x = 5$ and $2y = x - 5$.

Rewrite each equation in the form $y = mx + c$ and state whether the lines are parallel or not parallel.

Solution

Rearrange the first equation $\qquad\qquad 4y - 8x = 5$

Adding $8x$ to both sides gives $\qquad\qquad 4y = 8x + 5$

Dividing both sides by 4 gives $\qquad\qquad y = 2x + \dfrac{5}{4}$

Rearrange the second equation $\qquad\qquad 2y = x - 5$

Dividing both sides by 2 gives $\qquad\qquad y = \dfrac{1}{2}x - \dfrac{5}{2}$

The gradient of the line $y = 2x + \dfrac{5}{4}$ is 2.

The gradient of the line $y = \dfrac{1}{2}x - \dfrac{5}{2}$ is $\dfrac{1}{2}$.

As these gradients are different, the lines are *not* parallel.

Practice questions 3

1   For each pair of lines write the equation in the form $y = mx + c$ and state whether the lines are parallel or not parallel.

   a $y - 2x = 1$ and $2y = 4x + 7$ **b** $2y - 8x = 1$ and $8y = 2x - 16$
   c $6y - 3x = 4$ and $4y = 2x + 12$ **d** $x - 2y + 2 = 0$ and $2y - x - 4 = 0$

2   Look again at each pair of equations in question 1. For each equation write down the gradient m and the y-intercept c, and then sketch the two graphs on the same axes.

Practice exam questions

1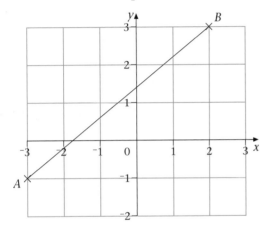

   Calculate the gradient of the straight line *AB*.

   [AQA (NEAB) 2001]

2   **a** On a copy of the grid, draw the line $y = 2x$.
   b The line $y = 2x$ crosses the line $x = -1$ at *P*.
   Give the coordinates of *P*.

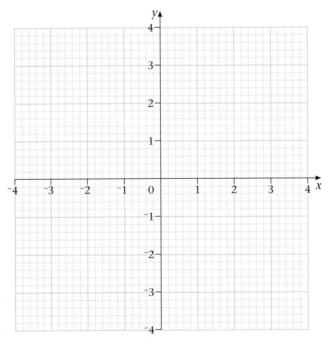

[AQA (SEG) 2000]

3 In an experiment, different weights are attached to a spring and the length
of the spring is measured each time.
The graph shows the results obtained each time.

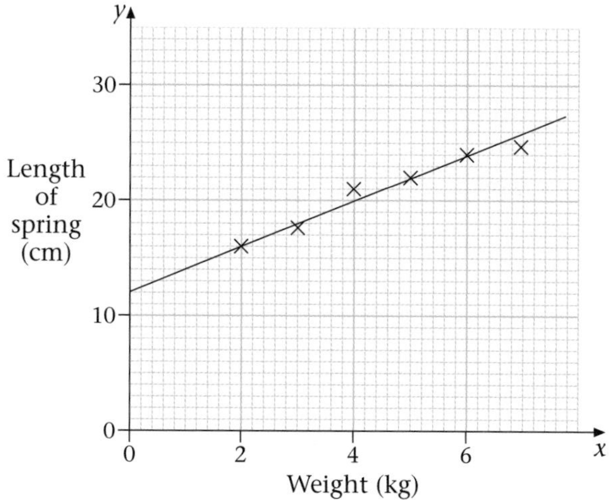

a Estimate the length of the spring when no weight is attached.
b Calculate the gradient of the line.
c Estimate the weight needed to *stretch* the spring by 20 cm.

[AQA (SEG) 1999]

16 Transformations

A **transformation** changes the position or size of a shape in various ways. You will need to know how to transform shapes using **translations**, **reflections**, **rotations** and **enlargements**.

You will also need to know and use **vector notation** when translating shapes.

Translations

A **translation** of a shape will move the shape horizontally and/or vertically.

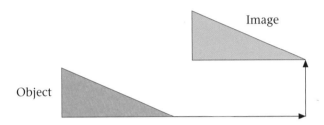

The new shape, the **image**, will always appear the same way up as the original shape, the **object**.

Vector notation

$\begin{pmatrix} a \\ b \end{pmatrix}$ means translate the shape *a* units in the *x*-direction and *b* units in the *y*-direction.

$\begin{pmatrix} a \\ b \end{pmatrix}$ is called a **translation vector** or a **column vector**.

If *a* is positive the movement will be to the right horizontally on your page. This is called the positive *x*-direction.

If *a* is negative the movement will be to the left horizontally on your page. This is called the negative *x*-direction.

If *b* is positive the movement will be upwards (vertically) on your page. This is called the positive *y*-direction.

If *b* is negative the movement will be downwards (vertically) on your page. This is called the negative *y*-direction.

Reminder
x-direction is horizontal, *y*-direction is vertical; like axes on a graph.

Describing translations

You may be asked to describe a translation shown on a graph. When asked to describe a translation there are **two** parts to the answer – the word **TRANSLATION** and **either the vector or both of the directions in words.**

Example 16.1

Describe the movement of a shape using the translation vector $\begin{pmatrix} 4 \\ -1 \end{pmatrix}$.

Solution

The shape will move 4 units in the positive *x*-direction (to the right) and 1 unit in the negative *y*-direction (downwards).

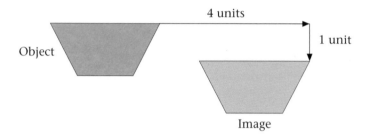

Example 16.2

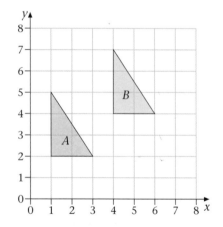

Describe the transformation from shape *A* to shape *B*.

Solution

Method A (using vector notation)

It is a translation of $\begin{pmatrix} 3 \\ 2 \end{pmatrix}$.

Method B (using words)

It is a translation of 3 units in the positive *x*-direction and 2 units in the positive *y*-direction.

Performing translations

You may also be asked to perform a translation of a shape on graph paper. Look at the vector notation. The top number tells you how many units the shape moves to the right or left. The bottom number tells you how many units the shape moves up or down.

Example 16.3

Transform the shape by a translation of $\begin{pmatrix} -3 \\ 5 \end{pmatrix}$.

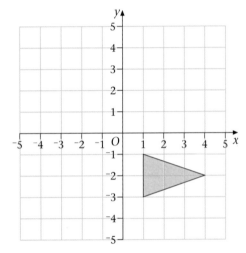

Solution

The -3 means move the shape 3 units in the negative x-direction and the 5 means move the shape 5 units in the positive y-direction.

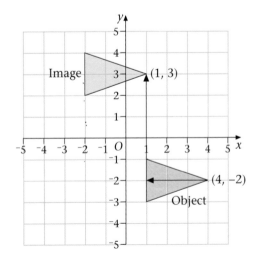

EXAMINER **TIP**

Check one vertex to see if the shape is in the correct position, e.g. the point $(4, -2)$ should move to $(4 - 3, -2 + 5) \rightarrow$ $(1, 3)$ as shown on the solution.

Practice questions 1

1   Describe the transformations of the shaded triangle *A* to:

   a triangle *B*   **b** triangle *C*   **c** triangle *D*   **d** triangle *E*.

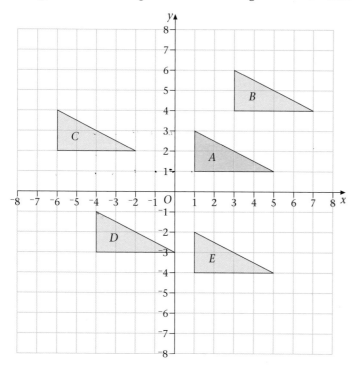

2   Translate the shaded shape using the following translation vectors.

   a $\begin{pmatrix} 4 \\ 2 \end{pmatrix}$ **b** $\begin{pmatrix} -6 \\ 3 \end{pmatrix}$ **c** $\begin{pmatrix} 2 \\ -5 \end{pmatrix}$ **d** $\begin{pmatrix} -6 \\ -8 \end{pmatrix}$

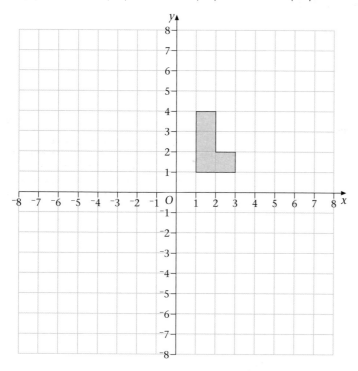

Reflections

A **reflection** is a mirror image of a shape in a given line. When a shape is reflected, the image is the same perpendicular distance from the mirror line as the object.

When asked to describe a reflection there are **two** parts to the answer – the word **REFLECTION** and **the equation of the mirror line**.

You will need to know how to reflect shapes in the following ways.

Reflection in the *x*-axis

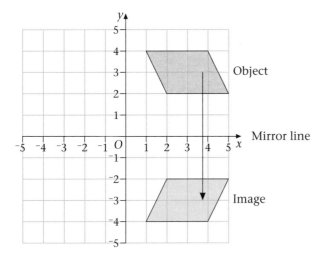

Reflection in the *y*-axis

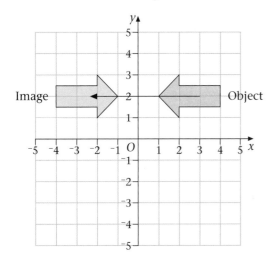

Reflection in the line $x = c$

Example 16.4

Reflect this shape in the line $x = -1$.

Solution

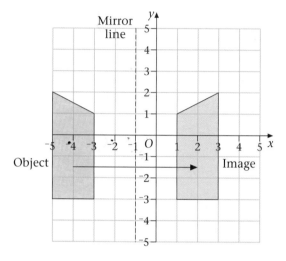

Reflection in the line $y = c$

Example 16.5

Reflect this shape in the line $y = 2$.

Solution

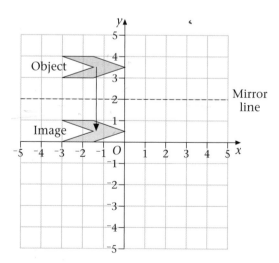

Reflection in the line *y = x*

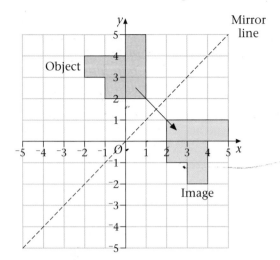

Practice questions 2

1 Reflect the shaded shape
in each of the lines
y = x, *y = −x*, *x* = 0 and *y* = 0.

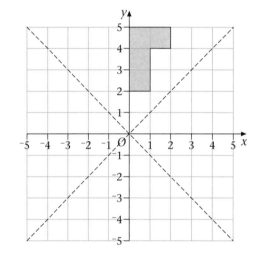

2 Describe fully the *single* transformation that moves the object *A* to the image *B* for each of the following.

a

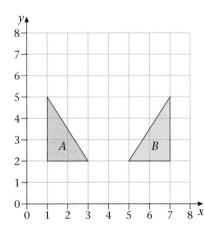

b

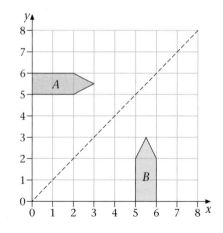

Rotations

A **rotation** is the turning of a shape about a fixed point, called the centre of rotation, through a particular angle in a clockwise or anticlockwise direction. You will need to know how to rotate shapes in the following ways.

When asked to describe a rotation there are **three** parts to the answer – the word **ROTATION**, the **angle with the direction of rotation** and the **centre of rotation**.

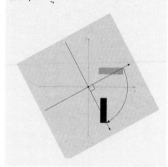

Rotation through 180° about (0, 0)

180° is the only angle where it is not necessary to state the direction.

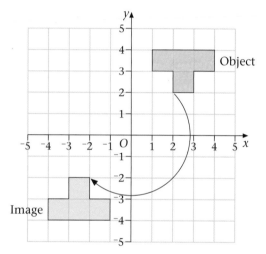

Rotation through 90° clockwise about (0, 0)

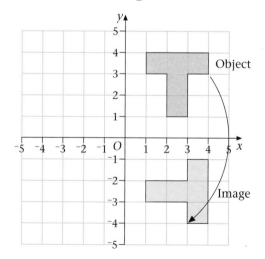

Rotation through 90° anticlockwise about (0, 0)

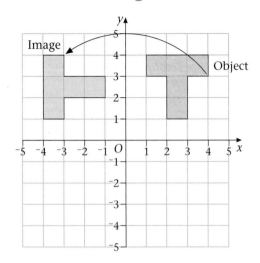

You will also need to rotate shapes about points other than the origin (0, 0).

Rotation through 90° clockwise about (−1, 1)

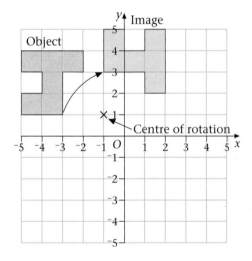

Example 16.16

Rotate the triangle *ABC* through 90° clockwise about (0, 0).

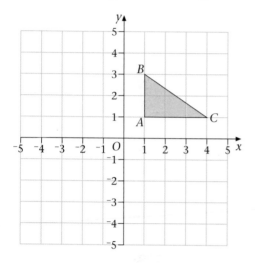

Solution

There are three parts to this transformation.

1 Rotation

2 90° clockwise

3 about the centre (0, 0)

It is useful to use tracing paper to help you here.

Trace the shape and the axes. Pivot the tracing paper about the centre of rotation (0, 0) through 90° clockwise in order to identify the position of the image.

Mark the position of the vertices of the image on your grid and complete the drawing of the image.

Rotating clockwise as shown in the diagram gives:

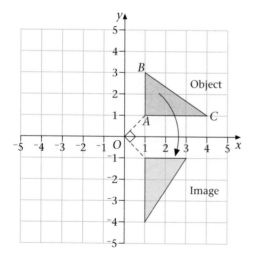

EXAMINER *TIP*

If you don't have any tracing paper, draw a line from any vertex on the object to the centre of rotation, then draw a line of the same length at 90°. Mark the image vertex at the end of the line. Repeat for the other vertices.

Example 16.7

Rotate the shaded shape through 180° about (2, –1).

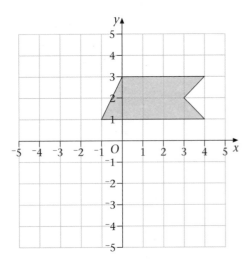

Solution

There are three parts to this transformation.

1 Rotation

2 180°

3 about the centre (2, –1)

It is useful to use tracing paper to help you here.

Trace the shape and the axes. Pivot the tracing paper about the centre of rotation (2, –1) through 180° in order to identify the position of the image.

Mark the position of the vertices of the image on your grid and complete the drawing of the image.

Rotating clockwise as shown in the diagram gives:

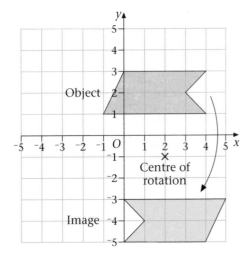

Example 16.8

The diagram opposite shows two identical shapes. Describe fully the single transformation that takes shape *A* to shape *B*.

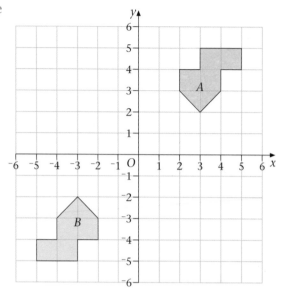

Solution

It is a rotation of 180° about centre (0, 0). You can use tracing paper to help you see the angle and centre of rotation.

Place the tracing paper over the grid and draw shape *A* on the paper. Now rotate the paper around holding down the paper in different positions with your pencil. You know you have found the centre of rotation when shape *A* on the tracing paper overlaps shape *B* on the grid.

To find the angle think about how far round you have rotated the shape *A* to get to shape *B*. As shape *B* is an upside down shape *A*, you can tell that it has been rotated 180°.

Practice questions 3

1　Rotate the shaded shape as follows:

　a　90° clockwise about (0, 0). Label it *A*.
　b　180° about (0, 0). Label it *B*.
　c　90° anticlockwise about (0, 0). Label it *C*.

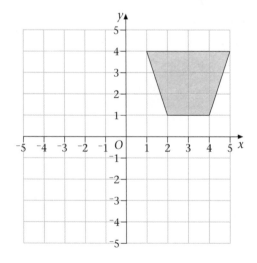

2　Rotate the shaded shape as follows:

　a　90° clockwise about (1, 1). Label it *A*.
　b　180° about (−1, 0). Label it *B*.
　c　90° anticlockwise about (−1, 1). Label it *C*.

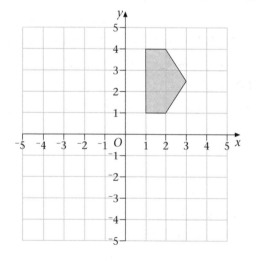

3   Describe the **single** transformation of:

   a   triangle *A* to triangle *B*
   b   triangle *B* to triangle *A*.

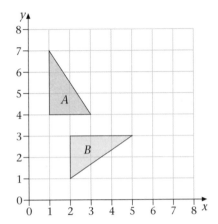

Enlargements

An **enlargement** is a transformation that increases or decreases the
size of a shape. The object and the image shapes will have sides
that are in proportion. For instance, if side *A* on the image is twice
as long as side *A* on the object, then side *B* on the image will be
twice as long as side *B* on the object as well.

The **scale factor** of an enlargement tells you how much larger or
smaller the lengths of the image are than the lengths of the object.

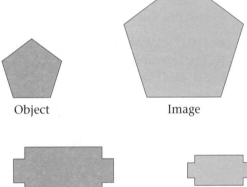

Object   Image

If the scale factor is greater than 1, the lengths of the image are
larger than the lengths of the object.

If the scale factor is less than 1, the lengths of the image are
smaller than the lengths of the object.

Object   Image

You will need to know how to enlarge a shape in the following
ways:

● on a grid by drawing the image in any position, given the scale factor of
   enlargement

● on a graph, drawing the image in an exact position, given a scale factor and
   a centre of enlargement.

When asked to describe an enlargement there are **three** parts to the answer –
the word **ENLARGEMENT**, the **scale factor of the enlargement** and the
centre of enlargement.

Example 16.9

Enlarge the shape on the grid by a scale factor 2.

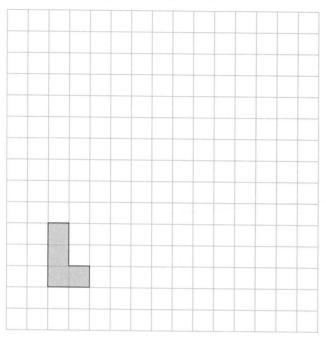

Solution

No centre of enlargement is given. This means that the enlarged shape (image) can be drawn anywhere on the grid.

A scale factor 2 means that the length of each side is multiplied by 2.

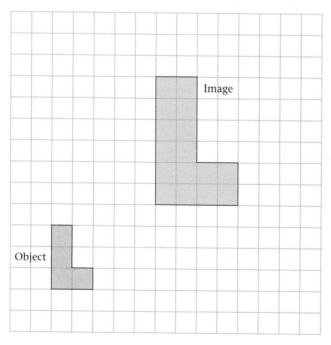

Example 16.10

Enlarge the triangle by scale factor 2 with centre of enlargement (0, 0).

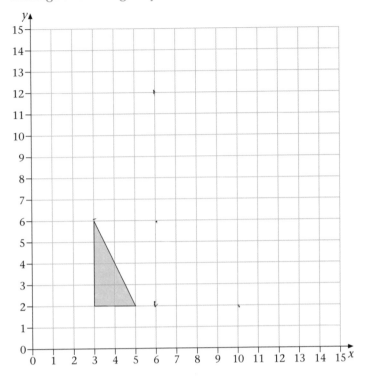

Solution

As the centre of enlargement is given there is only one correct position for the image.

In this case the distance of each point from the centre of enlargement is multiplied by 2 and the length of each side of the triangle will then be multiplied by 2.

The vertex of the triangle at (3, 2) will transform to the point (6, 4) and the other vertices will transform to the new positions as shown by the arrows.

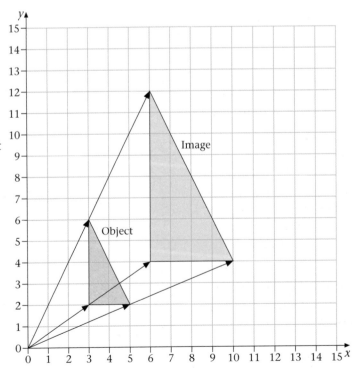

Example 16.11

Enlarge the shaded shape by scale factor $\frac{1}{3}$ with centre of enlargement (1, 2).

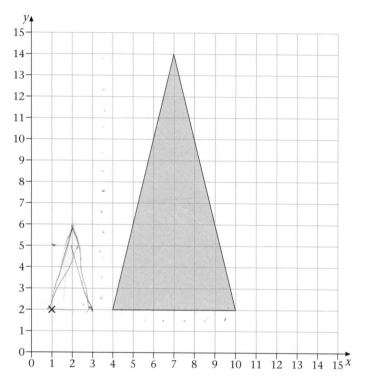

Solution

As the centre of enlargement is given, there is only one correct position for the image.

In this case the distance of each point from the centre of enlargement is multiplied by $\frac{1}{3}$ and the length of each side of the triangle will also be multiplied by $\frac{1}{3}$.

The distance between the vertex of the triangle at (4, 2) and the centre of enlargement is 3 horizontal units. This vertex on the image is $\frac{1}{3}$ the distance away from the centre of enlargement, which is 1 horizontal unit. So the vertex on the image is at (1 + 1, 2) which is (2, 2).

The vertex of the triangle at (7, 14) will transform to the point (3, 6) as its distance from the centre of enlargement is now $\frac{1}{3}$ its original distance from the centre of enlargement as shown by the arrows; similarly for the vertex at (10, 2).

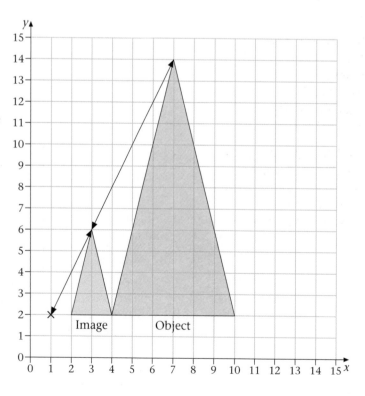

Example 16.12

Describe the *single* transformation that transforms the shape *A* to the shape *B* as shown.

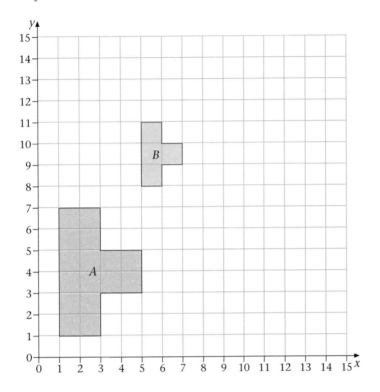

Solution

There are three parts to this transformation:

1 enlargement

2 scale factor

3 centre of enlargement.

As the size of the shape has changed it has to be an enlargement (even though the image is smaller).

As the lengths of the image shape are all half of the lengths of the object shape, the scale factor is $\frac{1}{2}$.

To find the centre of enlargement draw lines through corresponding points on the object and image. These lines will intersect at the centre of enlargement. Looking at the diagram the centre of enlargement is (9, 15).

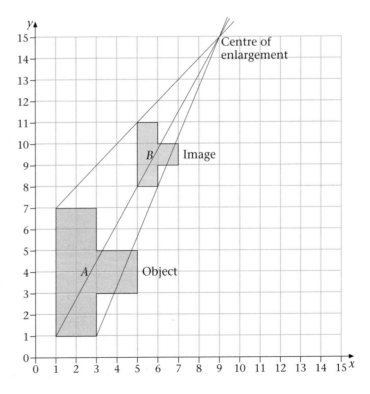

Example 16.13

Draw the following shapes on a grid: Square, Rectangle and L-shape.

Now enlarge each shape by scale factors of 2 and 3.

Compare the perimeter of each of the objects with each of their images.

What do you notice?

Solution

On a large grid draw the three shapes and enlarge each shape by scale factors of 2 and 3.

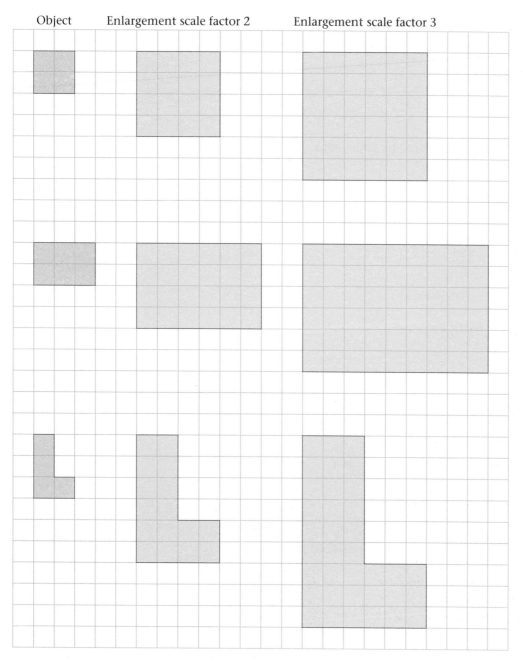

Now measure the perimeters of all the shapes. Compare each image with its object.

Shape	Perimeter		
	Object	Enlargement scale factor 2	Enlargement scale factor 3
Square	8	16	24
Rectangle	10	20	30
L-shape	10	20	30

When the scale factor is 2 the perimeter is twice as large. When the scale factor is 3 the perimeter is three times as large.

The perimeter is increased by the scale factor of the enlargement.

Practice questions 4

1   On a copy of the grid, enlarge the shape by scale factor 3.

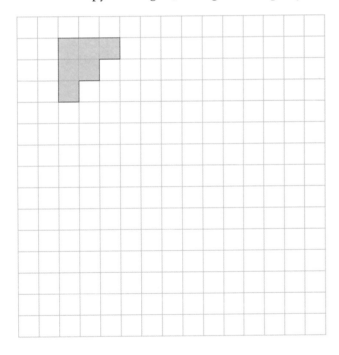

2 Describe fully the *single* transformation which maps shape *A* to shape *B*.

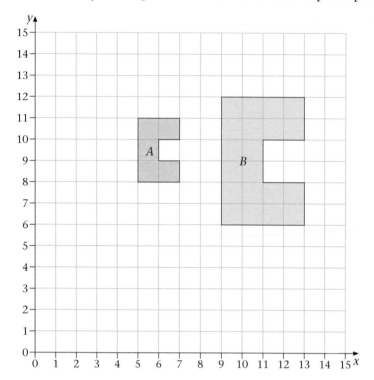

3 Enlarge the shape by a scale factor 2 with centre of enlargement (1, 2).

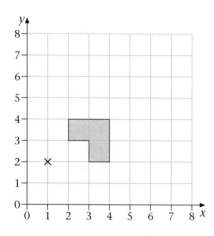

4 This shape has a perimeter of 3.45 m.

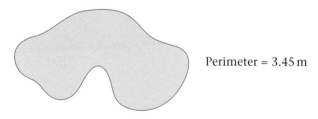

Perimeter = 3.45 m

Without using graph paper, write down the length of the perimeter after the shape has been enlarged by a scale factor of 3.

Combined transformations

You may be asked to perform two transformations, one straight after the other, on one shape. To do this you must perform the transformations in the order they are given.

Example 16.14

The diagram shows a shape, *A*.
A is reflected in the line *y* = *x* and then rotated anti-clockwise 90° about the point (0, –1).
Draw the final position of *A* after these transformations and label it *C*.

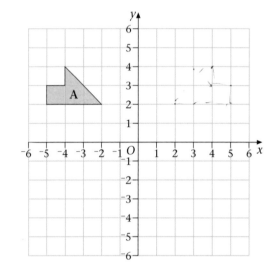

Solution

First reflect the shape *A* in the line *y* = *x*. Draw the line to help you. Each point on the object *A* is the same perpendicular distance from the mirror line as each point on the image *B*.

Now rotate the shape *B* 90° anti-clockwise about the point (0, –1). Mark the point on the grid. You can you use tracing paper to help you find the position of the rotated shape. Or you can draw construction lines to help you as is shown here.

Remember to label your shape *C*.

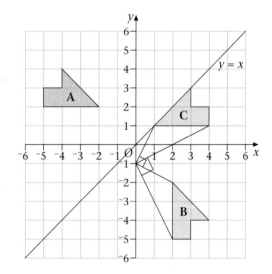

Planes of symmetry

Some *three-dimensional shapes* can be divided into two parts that are reflections of each other using a **plane of symmetry**. A cube has nine planes of symmetry. The diagrams show three of them. You will be expected to identify the number of planes of symmetry in a 3-D shape.

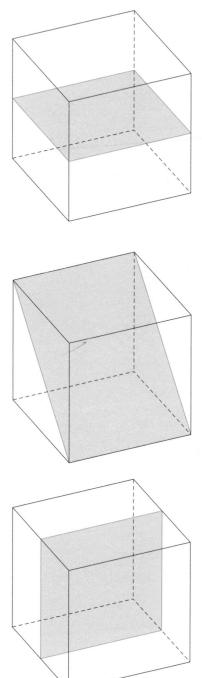

Practice question 5

1   Here are some 3-D shapes.
    How many planes of symmetry does each have?

a

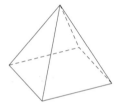

A cuboid with a rectangular cross-section.

b

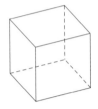

A cuboid with a square cross-section.

c

A square-based pyramid.

d

A solid T-shape.

An equilateral triangle-based pyramid.

e

A solid T-shape.

Practice exam questions

1

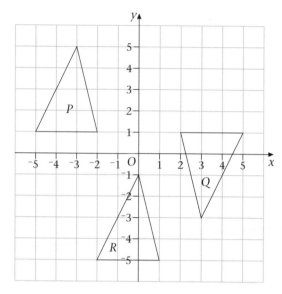

a Triangle *P* can be moved onto triangle *Q* using a rotation through 180°. Write down the coordinates of the centre of this rotation.

b Write down the vector for the translation which moves triangle *P* onto triangle *R*.

[AQA (SEG) 2001]

2

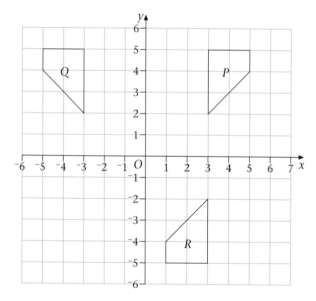

a Describe fully the single transformation which maps shape *P* onto shape *Q*.

b Describe fully the single transformation which maps shape *P* onto shape *R*.

c Shape *P* undergoes the translation vector $\begin{pmatrix} 2 \\ -5 \end{pmatrix}$.

Draw the new position of shape *P*.

[AQA (SEG) 2002]

3 Enlarge the shaded triangle by a scale factor of 3.
Use *P* as the centre of enlargement.

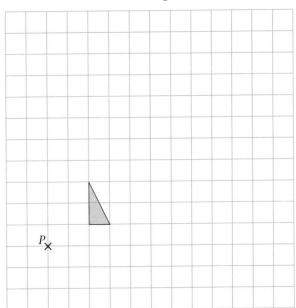

[AQA (NEAB) 2002]

4 The diagram shows three triangles *P*, *Q* and *R*.

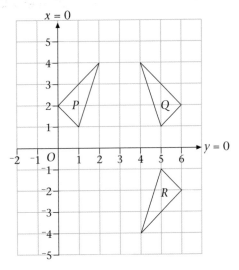

a Describe fully the single transformation which takes *P* onto *Q*.
b Describe fully the single transformation which takes *P* onto *R*.

[AQA (SEG) 1999]

5 The grid shows two congruent shapes, A and B.

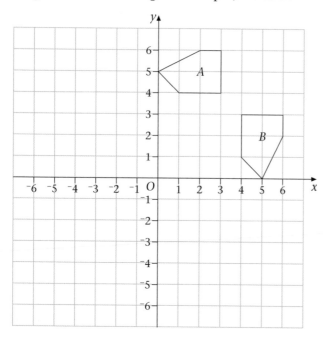

a On a copy of the grid, draw the rotation of shape A through 180° about O.

b Describe the *single* transformation of shape A to shape B.

[AQA (NEAB) 2002]

6

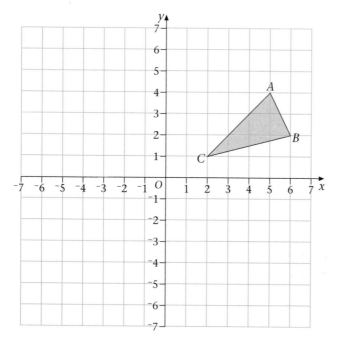

a Copy the diagram. Draw the reflection of the triangle ABC in the x-axis.

b i The point P(−4, 6) is reflected in the x-axis.
   What are its new coordinates?

  ii The point Q(25, 14) is reflected in the x-axis.
   What are its new coordinates?

[AQA (NEAB) 2001]

7   Find the coordinates of the reflection of the point (1, 4) in the line $y = -x$.
    (You may find the grid below useful.)

[AQA (NEAB) 2001]

8   The diagram shows the positions of triangles *P* and *Q*.

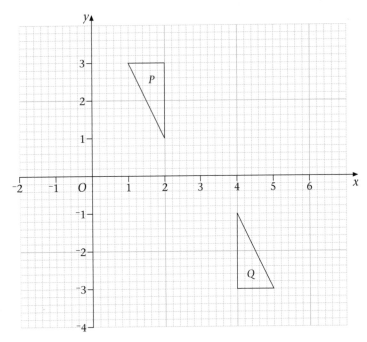

Describe fully the single transformation which maps *P* onto *Q*.   [AQA (SEG) 2000]

9 The diagram shows shapes Q and R which are transformations of shape P.

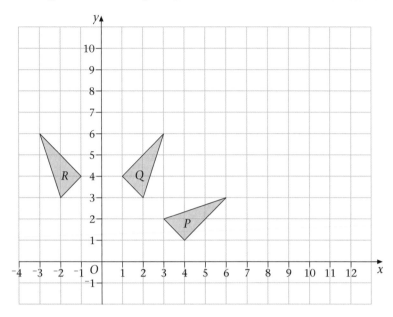

   a Describe fully the *single* transformation which takes P onto R.
   b Describe fully the *single* transformation which takes P onto Q.
   c On a copy of the diagram, draw an enlargement of shape P with scale
      factor 2, centre (3, 2).

[AQA (SEG) 1999]

10 On a copy of the diagram, enlarge the triangle with scale factor $\frac{1}{3}$, centre P.

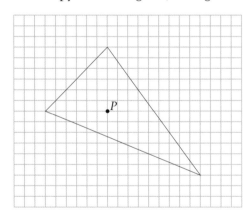

[AQA (SEG) 2000]

11 Triangle *ABC* has vertices *A*(6, 0), *B*(6, 9), *C*(9, 3).

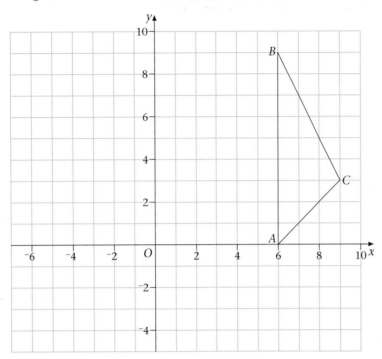

a On a copy of the diagram, rotate triangle *ABC* through 180° about the point (2, 4).
   Label the image triangle *R*.

b Enlarge triangle *ABC* by the scale factor $\frac{1}{3}$ from the centre of enlargement (3, 0).

   Label the image triangle *E*.

[AQA (NEAB) 2001]

12 *PQRS* is a parallelogram with vertices at *P*(1, 1), *R*(7, 8) and *S*(5, 3).

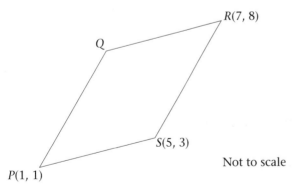

Not to scale

a Write down the coordinates of *Q*.

b The parallelogram is enlarged with scale factor $\frac{1}{2}$, centre *P*(1, 1).

   Write down the new coordinates of *S*.

[AQA (SEG) 1999]

13 The diagram shows two identical shapes *A* and *B*.
Describe fully the *single* transformation which takes shape *A* to shape *B*.

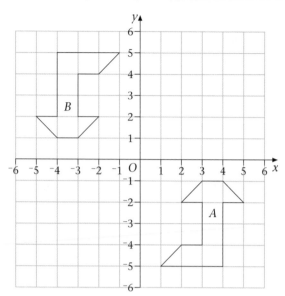

[AQA 2003]

14 The diagram shows two triangles, *C* and *D*.

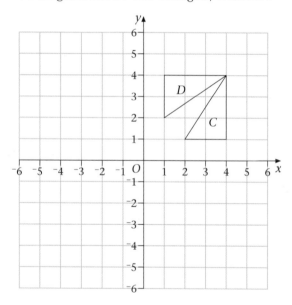

a Describe fully the *single* transformation which maps triangles *C* to triangle *D*.

b Triangle *C* is translated by the vector $\begin{pmatrix} -4 \\ -3 \end{pmatrix}$ and then rotated 90° anti-clockwise about the point (0, −2).

On a copy of the grid, draw the final position of triangle *C* after these transformations.

[AQA 2003]

17 Angle facts 2

Parallel lines

When a straight line crosses a pair of parallel lines several pairs of angles are equal.

You will need to know and use the following angle facts about parallel lines.

Alternate angles

Alternate angles are often referred to as Z-angles but you should use the correct name.

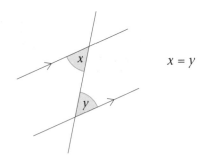

$x = y$

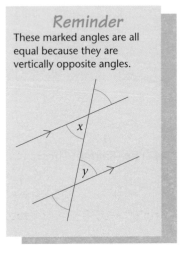

Reminder

These marked angles are all equal because they are vertically opposite angles.

Alternate angles are always equal.

Example 17.1

Write down the value of *y*.

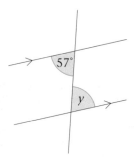

Solution

Angle *y* = 57° (alternate angles)

Corresponding angles

Corresponding angles are often referred to as
F-angles but you should use the correct name.
Corresponding angles are always equal.

$t = s$

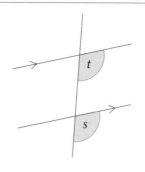

Example 17.2

Write down the value of t.

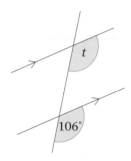

> EXAMINER **TIP**
>
> It is sometimes easier to work on the diagrams and then copy your answer onto the answer line.

Solution

Angle $t = 106°$ (corresponding angles)

Example 17.3

Work out the values of a and b giving reasons for your answers.

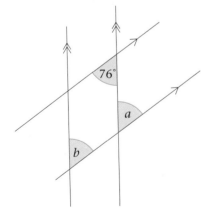

> *Reminder*
>
> Arrowheads on lines indicate that they are parallel. Here there is more than one set of parallel lines and so the number of arrowheads shows which lines are parallel.

Solution

$a = 76°$ (alternate angles are equal)
$b = 76°$ (a and b are corresponding angles)

Allied angles

Allied angles are inside a pair of
parallel lines as shown.

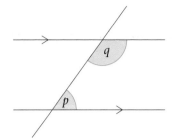

Allied angles add to 180°.
Angles p and q always add to 180°.

Example 17.4

Write down the value of q.

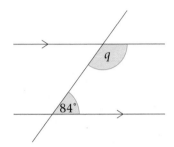

Solution

$84° + q = 180°$ (since 84° and q are allied angles)
$$q = 180° - 84°$$
$$= 96°$$

Example 17.5

Look at the diagram below.

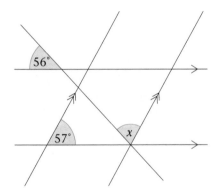

Find the size of x.

Solution

There are several ways you can work out the value of *x*. Here is one of them.

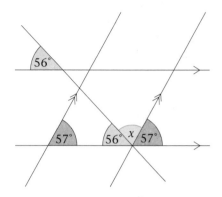

The angle to the right of *x* is 57°; it is a corresponding angle.

The angle to the left of *x* is 56°; it is a corresponding angle.

Then:
$56° + 57° + x = 180°$ (angles on a straight line)
$$x = 180° − 56° − 57°$$
$$x = 67°$$

Practice question

1 In each part, work out the missing angles giving reasons for your answers.

a
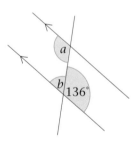

b
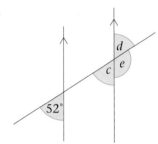

c
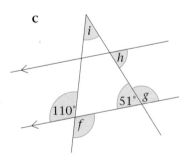

d
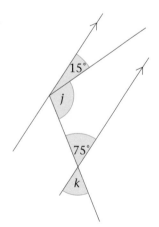

Practice exam questions

1   The diagram shows a rectangle with its diagonals drawn.

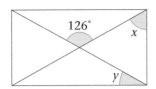

Not to scale

Work out the size of angle *x* and the size of angle *y*.   [AQA (SEG) 1999]

2   *BE* is parallel to *CD*.

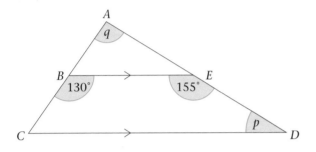

   a   Write down the size of angle *p*.
   b   Work out the size of angle *q*.   [AQA (NEAB) 2000]

3   *AB* is parallel to *CD*.

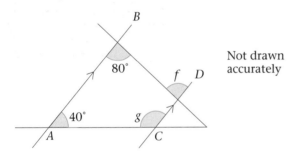

Not drawn
accurately

   a   Write down the size of the angle marked *f*.
   b   Calculate the size of the angle marked *g*.
   c   Mark clearly on a copy of the diagram another
           angle which is the same as the angle marked *g*.   [AQA (NEAB) 2001]

18 Factorising

Factorising linear equations

Factorising is the reverse process of multiplying out brackets or expanding brackets.

To **factorise** is to separate out common factors from an expression by using brackets.

> *Reminder*
> In Chapter 5, Expanding brackets, you multiplied out brackets, e.g.
> $3(x + 5) = 3x + 15$.

Factorising numbers

You can factorise by taking out the common factor of numbers in an expression.

Example 18.1

Factorise $6x + 18$.

Solution

Look at each term and see if there are any common factors.

In this example, 6 is the highest common factor of 6 and 18 and can be taken out of each term.

$6x + 18 = 6(x + 3)$

This is now fully factorised.

> *Reminder*
> The highest common factor (HCF) of two numbers is the highest number that is a factor of both numbers.

Example 18.2

Factorise $5x - 20$.

Solution

The common factor is 5, which can be taken out of each term.

$5x - 20 = 5(x - 4)$

This is now fully factorised.

Example 18.3

Factorise $2a - 8b$.

Solution

The common factor is 2, so $2a - 8b = 2(a - 4b)$.

Practice question 1

1   Factorise these expressions:

   a $3x + 12y$ **b** $4s - 24t$ **c** $15a + 3b$
   d $6x - 18y$ **e** $14p + 35q$ **f** $10e - 8f$

Factorising one common variable

When the terms in an expression have a common variable it can be factorised. The common factor may be a letter.

Example 18.4

Factorise $3c + 2cd$.

Solution

The only common factor in each term is the letter c. Factorising gives:
$3c + 2cd = c(3 + 2d)$

Example 18.5

Factorise $5x^2 - 18x$.

Solution

Both terms have a common factor of x.

$5x^2 - 18x = x(5x - 18)$

Practice question 2

1   Factorise:

   a $3p + 4pq$ **b** $5r - 8qr$ **c** $6s + 7st$
   d $x^2 - 16x$ **e** $3x^2 + 11x$ **f** $5x^2 + 12x$

Factorising one common variable and numbers

Expressions can have both numbers and letters as common factors and both can be factorised.

Example 18.6

Factorise completely $3x^2 - 15x$.

Solution

3 and x are common factors to both terms, so $3x$ is the HCF of these two terms.

$3x^2 - 15x = 3x(x - 5)$

EXAMINER **TIP**

If a question says factorise fully or factorise completely there will usually be more than one common factor.

EXAMINER **TIP**

Don't forget to check your answer by expanding the brackets, i.e. check that $3x(x - 5) = 3x^2 - 15x$.

Practice question 3

1 Factorise:

a $5p + 20pq$ **b** $2x^2 + 6x$ **c** $7st - 14t$
d $12a^2 + 3ab$ **e** $8r - 4qr$ **f** $9x^2 + 6x$

Factorising more than one common variable

Sometimes more than one letter can be factorised.

Example 18.7

Factorise $3x^2y - 12xy$.

Solution

3, x and y are common factors to both terms, so $3xy$ is the HCF.

$3x^2y - 12xy = 3xy(x - 4)$

With these more complicated expressions, you can factorise one term at a time. First take the common x-term out:

$3x^2y - 12xy = x(3xy - 12y)$

Then take the y-term out:

$x(3xy - 12y) = xy(3x - 12)$

Finally take out the number common factor of 3:

$xy(3x - 12) = 3xy(x - 4)$

Practice question 4

1 Factorise completely:

a $5xy - 12xy^2$ **b** $4pq + 20p^2q$ **c** $3rs^3 - 6r^2s$
d $cd^2 - c^2d$ **e** $5xyz + 15x^2yz^3$

Factorising quadratics

An expression of the form $ax^2 + bx + c$ where a, b and c are constants is called a quadratic expression, e.g. $x^2 + 10$, $x^2 + 4x - 3$, $x(x + 5)$ are all quadratic expressions.

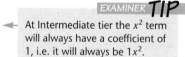

EXAMINER **TIP**

At Intermediate tier the x^2 term will always have a coefficient of 1, i.e. it will always be $1x^2$.

$x^2 + 7x + 12$ is a quadratic expression which can be factorised into two brackets.

x^2 is the first term, it is the squared term.

$+7x$ is the second term or the middle term.

$+12$ is the third term or the number term.

To factorise a quadratic expression you need to follow these steps.

Step 1

Start by drawing two pairs of empty brackets side by side (usually underneath the expression).

$x^2 + 7x + 12$
$(\quad)(\quad)$

Step 2

Consider the first term, the squared term. x^2 can only be made from x multiplied by x, so insert one x into each bracket.

$x^2 + 7x + 12$
$(x\quad)(x\quad)$

Step 3

Next consider the number term at the end of the expression. 12 has pairs of factors 1×12, 2×6 or 3×4. You have to choose the pair of factors which will add or subtract from each other to give the number 7 in the middle. Only 3 and 4 will add to 7, so insert 3 and 4 into the brackets.

$x^2 + 7x + 12$
$(x\quad 3)(x\quad 4)$

Step 4

Now consider the sign in front of the number term at the end of the expression, +12. In order to multiply the 3 and the 4 to get +12 the signs attached to the 3 and the 4 must be the same; either they are both positive signs or both negative signs. Before you decide which they are going to be you have to also consider the sign attached to the middle term +7x (positive). This tells you that the signs of the 3 and 4 must be both positive. Insert the signs in the brackets.

$x^2 + 7x + 12$
$(x + 3)(x + 4)$

Step 5

Always check your answer by multiplying out the brackets to see that you arrive at what you started with.

$$(x + 3)(x + 4) = x(x + 4) + 3(x + 4)$$
$$= x^2 + 4x + 3x + 12$$

Now you add the like terms: $4x + 3x = 7x$
giving $(x + 3)(x + 4) = x^2 + 7x + 12$ ✓. This expression has been factorised correctly.

Reminder
Look back to Chapter 5, Expanding brackets.

A checklist may be useful for working out the signs inside the brackets.

Sign of middle term	Sign of number term	Signs in the brackets
+	+	$(x +$ $)(x +$ $)$
−	+	$(x −$ $)(x −$ $)$
−	−	$(x +$ $)(x −$ $)$ or $(x −$ $)(x +$ $)$
+	−	$(x +$ $)(x −$ $)$ or $(x −$ $)(x +$ $)$

Example 18.8

Factorise $x^2 + 13x + 12$.

Solution

Step 1

Empty brackets

$x^2 + 13x + 12$

$($ $)($ $)$

Step 2

First term

$x^2 + 13x + 12$

$(x$ $)(x$ $)$

Step 3

Last term

$12 = 1 \times 12, 2 \times 6$ or 3×4

The middle number is 13, so choose 1×12 because these values add to 13.

$x^2 + 13x + 12$

$(x$ $1)(x$ $12)$

Step 4

The sign of the last term, +12, means both signs have to be the same: either both positive or both negative.

The Sign of the middle term, +13x, is positive, so both signs must be positive.

$x^2 + 13x + 12$

$(x + 1)(x + 12)$

Step 5

Check that you get back to where you started by expanding the brackets.

$(x + 1)(x + 12) = x^2 + 12x + 1x + 12$

$\qquad\qquad\qquad = x^2 + 13x + 12$ ✓, so it is factorised correctly.

Example 18.9

Factorise $x^2 - 8x + 12$.

Solution

Step 1
Empty brackets
$x^2 - 8x + 12$
(   )(   )

Step 2
First term
$x^2 - 8x + 12$
$(x \quad)(x \quad)$

Step 3
Last term

$12 = 1 \times 12, 2 \times 6$ or 3×4

Middle number is 8, so choose 2×6 because these values add to 8.

$x^2 - 8x + 12$
$(x \quad 2)(x \quad 6)$

Step 4
Sign of last term, +12, means both signs have to be the same: either both positive or both negative.

Sign of middle term, $-8x$, is negative, so both signs must be negative.

$x^2 - 8x + 12$
$(x - 2)(x - 6)$

Step 5
Check by expanding brackets that you get back to where you started.

$$(x - 2)(x - 6) = x^2 - 6x - 2x + 12$$
$$= x^2 - 8x + 12 \checkmark$$

Example 18.10

Factorise $x^2 + 8x - 20$.

Solution

Step 1
Empty brackets
$x^2 + 8x - 20$
$(\quad)(\quad)$

Step 2
First term
$x^2 + 8x - 20$
$(x\quad)(x\quad)$

Step 3
Last term

$20 = 1 \times 20, 2 \times 10$ or 4×5

Middle number is 8, so choose 2×10 because the difference between them is 8.

$x^2 + 8x - 20$
$(x\quad 2)(x\quad 10)$

Step 4
Sign of last term, -20, is negative which means the signs have to be different, one will be positive and the other will be negative.

Sign of middle term, $+8x$, is positive, so the larger number (10) must be positive.

$x^2 + 8x + 12$
$(x - 2)(x + 10)$

Step 5
Check by expanding brackets that you get back to where you started.

$(x - 2)(x + 10) = x^2 + 10x - 2x - 20$
$ = x^2 + 8x - 20 \checkmark$

Example 18.11

Factorise $x^2 - 4x - 5$.

Solution

Step 1
Empty brackets
$x^2 - 4x - 5$
$(\quad)(\quad)$

Step 2
First term
$x^2 - 4x - 5$
$(x\quad)(x\quad)$

Step 3

Last term

$5 = 1 \times 5$ so you have to choose these factors.

$x^2 - 4x - 5$

$(x \quad 1)(x \quad 5)$

Step 4

Sign of last term, -5, is negative which means the signs have to be different: one will be positive and the other will be negative.

Sign of middle term, $-4x$, is negative so the larger number (5) must be negative.

$x^2 - 4x - 5$

$(x + 1)(x - 5)$

Step 5

Check by expanding brackets that you get back to where you started.

$$(x + 1)(x - 5) = x^2 - 5x + 1x - 5$$
$$= x^2 - 4x - 5 \checkmark$$

Practice questions 5

1 Factorise:

    a $x^2 + 3x + 2$

    b $x^2 + 4x + 3$

    c $x^2 + 8x + 15$

    d $x^2 + 7x + 6$

    e $x^2 - 10x + 24$

    f $x^2 - 5x + 6$

2 Factorise:

    a $x^2 + 2x - 3$

    b $x^2 + 5x - 6$

    c $x^2 + 3x - 10$

    d $x^2 + 7x - 18$

    e $x^2 + 5x - 24$

3 Factorise:

    a $x^2 - 2x - 3$

    b $x^2 - 5x - 14$

    c $x^2 - 3x - 18$

    d $x^2 - 4x - 12$

    e $x^2 - 7x - 30$

4 Factorise:

    a $y^2 - 2y - 15$

    b $p^2 + 3p - 28$

    c $t^2 + 9t + 18$

    d $z^2 - 4z - 21$

Perfect squares

$(x - 3)^2$ is called a **perfect square** because it is $(x - 3)$ squared.

When you expand $(x - 3)^2$ you obtain $x^2 - 6x + 9$ which is a quadratic expression which is also called a perfect square.

Sometimes you are given a perfect square as a quadratic expression and asked to factorise it.

Example 18.12

Factorise $x^2 - 8x + 16$.

Solution

$$x^2 - 8x + 16 = (x - 4)(x - 4)$$
$$= (x - 4)^2$$

Practice question 6

1   Factorise the following expressions.

   a $x^2 + 4x + 4$

   b $x^2 + 6x + 9$

   c $x^2 + 10x + 25$

   d $x^2 - 4x + 4$

   e $x^2 - 8x + 16$

   f $x^2 - 12x + 36$

Difference of two squares

A **difference of two squares** (shortened to DOTS) is a squared term subtracted from another squared term, e.g. $x^2 - y^2$ or $x^2 - 25$. These are special types of quadratic expression and they can be factorised very quickly. In general, $x^2 - y^2 = (x + y)(x - y)$.

 EXAMINER **TIP**

Look out for DOTS, they are often tested in the exam papers.

Example 18.13

Factorise $x^2 - 100$.

Solution

Step 1
Empty brackets
$x^2 - 100$
(   )(   )

Step 2
First term
$x^2 - 100$
$(x$ $)(x$ $)$

Step 3
This is a DOTS; the last term is a square term: $100 = 10 \times 10$.

And because there is no middle term in the quadratic, insert $+10$ and -10 straight away into the brackets:
$x^2 - 100$
$(x + 10)(x - 10)$
Check answer $(x + 10)(x - 10) = x^2 - 10x + 10x - 100$
$$= x^2 - 100 \checkmark$$

Practice question 7

1   Factorise:

    a $x^2 - 1$ **b** $y^2 - 16$ **c** $z^2 - 225$

    d $t^2 - 64$ **e** $r^2 - 81$ **f** $s^2 - 121$

    g $36 - x^2$ **h** $25 - p^2$ **i** $1 - q^2$

Using the difference of two squares

DOTS are useful to help you calculate differences between square numbers without a calculator.

Consider how you would find the value of $91^2 - 9^2$ without a calculator.

$$91^2 - 9^2$$

This can be factorised: $= (91 - 9)(91 + 9)$

$= 82 \times 100$

$= 8200$

> **Reminder**
> You are expected to know all square numbers from 1^2 to 15^2.

Practice questions 8

1   Work out, without a calculator:

    a $51^2 - 49^2$ **b** $90^2 - 10^2$ **c** $28^2 - 12^2$

2   Rewrite each expression as the difference of two squares and factorise it to find the value of each expression without using a calculator.

    a $98^2 - 4$ **b** $89^2 - 121$ **c** $47^2 - 9$

Practice exam questions

1   Factorise completely:

    a $6a + 10ab - 8a^2b$

    b $a^2 - 9$ [AQA (NEAB) 2002]

2   Factorise completely $3x^2 - 6x$.   [AQA (NEAB) 1998]

3   Factorise completely $5x^2y + 15xy^3$.   [AQA (NEAB) 1999]

4   Factorise $x^2 - 10x + 21$.   [AQA (NEAB) 2002]

5   Factorise fully $3a - 27a^2$.   [AQA (NEAB) 2001]

6   Factorise:

    a $3pq - 6r$

    b $c^2 - 9c + 20$ [AQA (SEG) 2001]

7   Factorise $2x^2 + 4x$.   [AQA (SEG) 2000]

19 Cancelling common factors

You may be asked to simplify expressions that are written as fractions and have common factors.

Example 19.1

Simplify $\dfrac{x^2y}{xy^3}$.

Solution

Look at one letter at a time.

Looking at the x-terms, you have:

$\dfrac{x^2}{x}$ which means $\dfrac{x \times \cancel{x}}{\cancel{x}}$

This cancels down to leave an x in the numerator only.

Now look at the y-terms, you have:

$\dfrac{y}{y^3} = \dfrac{\overset{1}{\cancel{y}}}{y \times y \times \cancel{y}}$

This cancels down to leave 1 in the numerator and $y \times y$ in the denominator.

Now combine the parts:

$$\dfrac{x^2y}{xy^3} = \dfrac{x \times 1}{y \times y}$$

giving $\dfrac{x}{y \times y} = \dfrac{x}{y^2}$

Example 19.2

Simplify $\dfrac{2(x+1)^2}{(x+1)}$.

Solution

Write out the bracket squared in the numerator as the product of two identical brackets.

$$\dfrac{2(x+1)^2}{(x+1)} = \dfrac{2(x+1)(\cancel{x+1})}{(\cancel{x+1})}$$

One of the brackets in the numerator is identical to the bracket in the denominator and because all terms are being multiplied they cancel out.

Then $\dfrac{2(x+1)^2}{(x+1)} = 2(x+1)$ which is as simple as it can be.

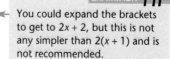

EXAMINER **TIP**

You could expand the brackets to get to $2x + 2$, but this is not any simpler than $2(x + 1)$ and is not recommended.

Example 19.3

Simplify $\dfrac{2(a^2 - 9)}{(a + 3)}$.

Solution

The bracket in the numerator is a difference of two squares and can be factorised.

> *Reminder*
> Look back at Chapter 18, Factorising, for the difference of two squares.

$$\frac{2(a^2 - 9)}{(a + 3)} = \frac{2(a - 3)(a + 3)}{(a + 3)} = \frac{2(a - 3)\cancel{(a + 3)}}{\cancel{(a + 3)}}$$

Since the bracket $(a + 3)$ is common to the numerator and denominator and all terms are being multiplied they will cancel out.

So $\dfrac{2(a^2 - 9)}{(a + 3)} = \dfrac{2(a - 3)(a + 3)}{(a + 3)} = 2(a - 3)$.

Example 19.4

Simplify $\dfrac{3(b^2 - 4)}{9(b - 2)}$.

Solution

The bracket in the numerator is a difference of two squares and can be factorised.

$$\frac{3(b^2 - 4)}{9(b - 2)} = \frac{^1\cancel{3}(b - 2)(b + 2)}{_3\cancel{9}(b - 2)}$$

The bracket $(b - 2)$ cancels out and the 3 and the 9 will cancel leaving $1(b + 2)$ in the numerator and 3 in the denominator.

EXAMINER **TIP** ← $1(b + 2)$ is the same as $(b + 2)$.

So $\dfrac{3(b^2 - 4)}{9(b - 2)} = \dfrac{(b + 2)}{3}$.

Practice question

1 Simplify the following expressions.

    a $\dfrac{a^3}{a^2}$ **b** $\dfrac{y^5}{y^3}$ **c** $\dfrac{ab^2}{a^2 b}$ **d** $\dfrac{x(x - 2)}{(x - 2)^2}$

    e $\dfrac{2y(y + 1)^2}{4(y + 1)}$ **f** $\dfrac{3(x^2 - 1)}{(x + 1)}$ **g** $\dfrac{9x^2 y^3}{3x^3 y}$

Practice exam question

1 Simplify $\dfrac{a^6 c^4}{a^2 c^5}$.

[AQA (NEAB) 1998]

20 Solving quadratic equations

An equation involving a quadratic expression is called a quadratic equation.

$x^2 + 7x + 12 = 0$ is a quadratic equation which can be solved by finding any values of x which will satisfy the equation.

The x-values that make the equation correct are called the **solutions** of the equation.

Quadratic equations can be solved by factorisation.

> **Reminder**
> An equation involving a squared variable is called a quadratic equation.

> *EXAMINER* **TIP**
> Read questions carefully so you know whether you are being asked to solve or factorise equations.

Example 20.1

Use factorisation to solve $x^2 + 9x + 8 = 0$.

Solution

$$x^2 + 9x + 8 = 0$$
Factorise the LHS $(x + 1)(x + 8) = 0$

If $(x + 1) \times (x + 8) = 0$, either $x + 1$ must be zero or $x + 8$ must be zero.

If $(x + 1) = 0$
then $x + 1 = 0$ because brackets are not needed
so $x = -1$, this is one solution.

If $(x + 8) = 0$
then $x + 8 = 0$
so $x = -8$, this is the other solution.

Factorisation is an efficient method of solving quadratic equations.

> **Reminder**
> You should already know how to factorise a quadratic expression, see Chapter 18, Factorising.

> *EXAMINER* **TIP**
> LHS means the left-hand side.

> *EXAMINER* **TIP**
> All the quadratic equations that you are asked to solve will factorise and can be solved using this method.

Practice questions 1

1 Solve these equations:
   a $x^2 + 7x + 10 = 0$
   b $x^2 + 2x - 8 = 0$
   c $x^2 + 5x - 6 = 0$
   d $x^2 + 5x + 6 = 0$
   e $x^2 - 6x - 27 = 0$
   f $x^2 - 9x - 10 = 0$

2 Solve these equations:
   a $y^2 + 9y - 10 = 0$
   b $p^2 - 8p - 9 = 0$
   c $t^2 + 3t - 4 = 0$
   d $s^2 - 2s - 3 = 0$
   e $z^2 - 8z + 12 = 0$

Sometimes you may be given quadratic equations that have no number term, e.g. $x^2 - 9x = 0$. This is still a quadratic equation and is best solved by factorisation.

$x^2 - 9x = 0$

$x(x - 9) = 0$

This gives $x = 0$ or $x - 9 = 0$, so the two solutions are $x = 0$ and $x = 9$.

EXAMINER **TIP**

Do not divide both sides of the equation by x. This would lose the solution $x = 0$.

Practice questions 2

1   Solve the following equations:

   a $x^2 - 8x = 0$
   b $x^2 + 2x = 0$
   c $x^2 - 5x = 0$
   d $x^2 + 3x = 0$
   e $x^2 - x = 0$

2   Solve the following equations:

   a $y^2 - 12y = 0$
   b $p^2 - 7p = 0$
   c $s^2 + 6s = 0$
   d $t^2 + 23t = 0$
   e $z^2 - 15z = 0$

Solving difference of two squares equations

You may be asked to solve a quadratic equation involving the difference of two squares.

Reminder

Look at the DOTS section in Chapter 18, Factorising.

Example 20.2

Solve $x^2 - 49 = 0$.

Solution

Method A (factorisation)

$x^2 - 49 = 0$

The LHS is a difference of two squares and can be factorised.

$(x + 7)(x - 7) = 0$

As before you have two terms multiplied together giving an answer of zero: the bracket $x + 7 = 0$ or the bracket $x - 7 = 0$.

This gives the two solutions $x = -7$ or $x = +7$.

Reminder

Look up how to factorise DOTS. See Chapter 18, Factorising.

Reminder

$x = \pm 7$ means $x = +7$ or $x = -7$.

Method B (rearranging)

$x^2 - 49 = 0$

Add 49 to both sides:

$x^2 - 49 + 49 = 0 + 49$

   giving $x^2 = 49$

then take the square root of both sides:

$\sqrt{x^2} = \sqrt{49}$

   $x = \pm 7$

EXAMINER **TIP**

Candidates who use this method often forget that there are two answers, +7 and −7. You will lose a mark for missing the −7 answer.

Practice question 3

1   Solve the following equations:

   a $x^2 - 81 = 0$ **b** $x^2 - 16 = 0$ **c** $x^2 - 144 = 0$
   d $x^2 - 169 = 0$ **e** $x^2 = 225$

Forming and solving quadratic equations

You may be required to form a quadratic equation from a given situation and then solve it to answer a question about the situation.

Example 20.3

A rectangle has sides of length $(x + 5)$ cm and $(x - 2)$ cm.

The area of the rectangle is 18 cm^2.

a Use the given information to form a quadratic equation and solve it to find the value of x.

b Write down the lengths of the sides of the rectangle.

Solution

It is always useful to draw a diagram and insert all the given information onto it.
To calculate the area of a rectangle you multiply the lengths of the sides together.

$(x - 2)$ cm

$(x + 5)$ cm

Area = 18 cm^2

This gives $(x + 5) \times (x - 2)$ which is the same as $(x + 5)(x - 2)$.

You are told that the area of this rectangle is 18 cm^2, so you can write $(x + 5)(x - 2) = 18$.
Expand the brackets on the LHS to get $x^2 + 3x - 10 = 18$.
This is a quadratic equation but it does not equal zero.

Rearrange the equation by subtracting 18 from both sides.
$x^2 + 3x - 10 - 18 = 18 - 18$
   $x^2 + 3x - 28 = 0$

> **Reminder**
> You can only solve quadratic equations by factorising when the equation is equal to zero.

This quadratic equation is now equal to zero and can be factorised.
   $(x + 7)(x - 4) = 0$
So $x = -7$ or $x = +4$
Although the quadratic equation has two solutions, x cannot be -7 as this would make the lengths of the sides both negative and it is impossible to have a negative length.

So the value of x has to be $+4$.

Substituting $x = +4$ into the lengths of the sides of the rectangle you have
$(x + 5) = (4 + 5) = 9$
and $(x - 2) = (4 - 2) = 2$.

So the lengths of the sides of the rectangle are 9 cm and 2 cm.

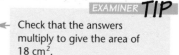
EXAMINER TIP
Check that the answers multiply to give the area of 18 cm^2.

Example 20.4

These two rectangles have the same area.

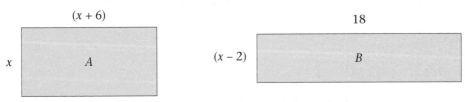

a Form a quadratic equation and solve it.

b Write down the lengths of the sides of rectangle A.

c Show that the perimeter of rectangle B is 8 cm more than the perimeter of rectangle A.

Solution

a Rectangle A has area $x(x + 6)$.
Rectangle B has area $18(x - 2)$.
The area of rectangle A = the area of rectangle B.
So $\qquad\qquad\qquad x(x + 6) = 18(x - 2)$
then $\qquad\qquad\qquad x^2 + 6x = 18x - 36$
rearranging $\qquad x^2 + 6x - 18x + 36 = 0$
$\qquad\qquad\qquad\quad x^2 - 12x + 36 = 0$
$\qquad\qquad\qquad\quad (x - 6)(x - 6) = 0$
$\qquad\qquad\qquad\qquad\qquad x = +6$

b Rectangle A has sides 6 cm and $(x + 6) = 12$ cm.

c Since you have found x you can write down the perimeter of each rectangle.
Rectangle A has perimeter $6 + 12 + 6 + 12 = 36$ cm.
Rectangle B has perimeter $4 + 18 + 4 + 18 = 44$ cm.
Hence rectangle B has a perimeter which is $44 - 36 = 8$ cm more than the perimeter of rectangle A.

Practice questions 4

1 A rectangle has sides of length $(x + 6)$ cm and $(x + 2)$ cm.
The area of the rectangle is 77 cm^2.

   a Use the given information to form a quadratic equation and solve it to find the value of x.

   b Write down the lengths of the sides of the rectangle.

2 A square has sides of length $(x + 3)$ cm.
The area of the square is 49 cm^2.

   a Use the given information to form a quadratic equation and solve it to find the value of x.

   b Write down the length of the side of the square.

3 The two rectangles have the same area.

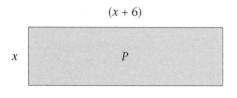

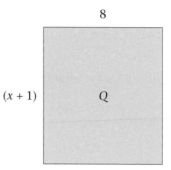

   a Form a quadratic equation and solve it.

   b Write down the lengths of the sides of rectangle P.

   c Show that the difference between the perimeters of these two rectangles is 2 cm.

Practice exam questions

1 The diagram shows a shape, in which all the corners are right angles.
The area of the shape is 48 cm^2.

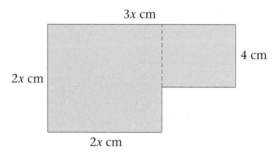

Not to scale

   a Form an equation, in terms of x, for the area of the shape.
Show that it can be simplified to $x^2 + x - 12 = 0$.

   b By solving the equation $x^2 + x - 12 = 0$, calculate the value of x.   [AQA (SEG) 1999]

2 **a** Work out the value of $x^2 - x - 12$ when $x = 4$.

   b Factorise $x^2 - x - 12$.

   c Hence write down both solutions of $x^2 - x - 12 = 0$.   [AQA (NEAB) 2001]

3 Solve $x^2 + 8x + 12 = 0$.
Do not use a trial and improvement method.   [AQA (NEAB) 1999]

4 **a** Factorise the expression $x^2 - 3x - 28$.

   b Hence solve the equation $x^2 - 3x - 28 = 0$.   [AQA (NEAB) 2001]

21 Quadratic graphs

A **quadratic graph** is one whose equation is a quadratic expression, e.g. $y = x^2$, $y = x^2 + 5$, $y = 2x^2 + 5$, $y = 3x^2 + x - 1$.

Quadratic graphs are curved graphs. They have a special shape called a parabola. It is useful to remember the following facts about quadratic graphs:

● If the equation of the graph has a positive x^2 term then the parabola will always be $\lor$-shaped.

● If the equation of the graph has a negative x^2 term then the parabola will always be $\land$-shaped.

● All quadratic graphs have a line of symmetry.

Plotting quadratic graphs

To plot a graph you need to calculate all of the pairs of coordinates that the graph passes through.
It is usual to set up a table of coordinates.

Example 21.1

a Complete the table of coordinates for the graph $y = x^2 + 2$.
b Use the table of coordinates to plot the graph of $y = x^2 + 2$.

EXAMINER TIP
You will always be given the graph paper on which to draw the graph.

x	−3	−2	−1	0	1	2	3
x^2	9	4		0			
+2	+2	+2	+2	+2	+2	+2	+2
y	11	6		2			

Solution

a In the top row of the table are the x-coordinates, these are the values along the x-axis.

The second row of the table contains the values of x^2, e.g. when $x = -3$, $x^2 = 9$.

The third row of the table is the +2 from the equation $y = x^2 + 2$. You add the +2 to every x^2 value to get the y-coordinate.

The completed table looks like this:

x	−3	−2	−1	0	1	2	3
x^2	9	4	1	0	1	4	9
+2	+2	+2	+2	+2	+2	+2	+2
y	11	6	3	2	3	6	11

b If the axes are not given on the graph paper then you will have to draw your own axes.

The *x*-axis starts at −3 and goes across to +3.

From the table of coordinates, the *y*-coordinates start at 2 and go up to 11.

When drawing graphs it is usual to have the origin (0, 0) shown on the axes.

It is also common to have the same size scale on both of the axes.

Once the axes have been marked, you plot the pairs of coordinates as shown in the table of coordinates.

The first point has coordinates (−3, 11), the next point has coordinates (−2, 6) and so on.

Ensure you plot the points accurately, and then join them with a smooth curve.

Finally label the graph.

Notice that the graph is symmetrical about the *y*-axis and that the *y*-coordinates in the table of coordinates are symmetrical too.

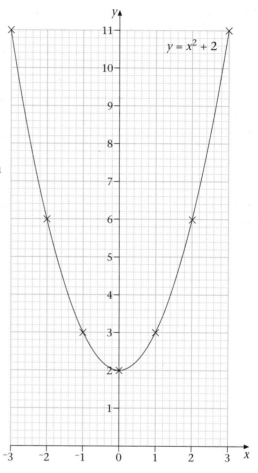

Example 21.2

a Complete the table of coordinates for the graph $y = x^2 + 3x - 4$.

x	−3	−2	−1	0	1	2	3
x^2		4		0			9
$+3x$		−6		0			9
−4	−4	−4		−4			−4
y		−6					14

b Use the table of coordinates to plot the graph of $y = x^2 + 3x - 4$.

141

Solution

a Complete the table as in Example 21.1.

The $+3x$ row means that you multiply the value of x by $+3$ and insert these values into the middle row of the table.

The completed table looks like this:

x	−3	−2	−1	0	1	2	3
x^2	9	4	1	0	1	4	9
$+3x$	−9	−6	−3	0	3	6	9
-4	−4	−4	−4	−4	−4	−4	−4
y	−4	−6	−6	−4	0	6	14

EXAMINER *TIP*

◄ In the exam you may be asked to complete a table showing only the *x*- and *y*-coordinates for a particular graph.

b Draw and label the axes. Note that the *y*-coordinates range from −6 to +14 but allow for the possibility that the graph may go below the lowest point.

Plot the points (−3, −4), (−2, −6), (−1, −6), etc.

Join the points with a smooth curve and label the graph.

Because the curve is a smooth curve it actually goes below the −6 on the *y*-axis. You should try to make your graphs smooth curves like this one as shown below.

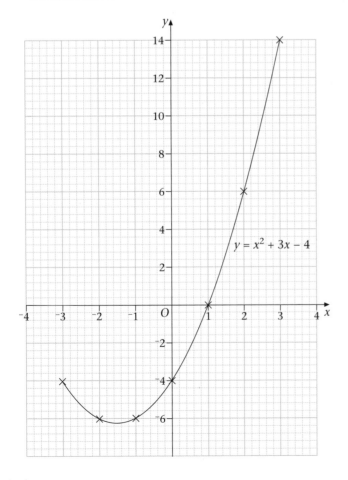

$y = x^2 + 3x - 4$

Using a graph to solve an equation

Occasionally the question may give you a completed table of coordinates and ask you to draw the graph and then use the graph to find solutions to equations.

Example 21.3

a Complete this table of coordinates and use it to draw the graph of $y = x^2 + x - 6$.

x	-4	-3	-2	-1	0	1	2	3
y	6	0	-4					

b Use your graph to find the solutions of $x^2 + x - 6 = 0$.

Solution

a

x	-4	-3	-2	-1	0	1	2	3
y	6	0	-4	-6	-6	-4	0	6

b To find the solutions of the quadratic equation $x^2 + x - 6 = 0$, find the x-coordinates where the graph crosses the x-axis (in other words where $y = 0$).

These are the points circled on the graph.

The graph crosses the x-axis at -3 and $+2$ and so these are the solutions.

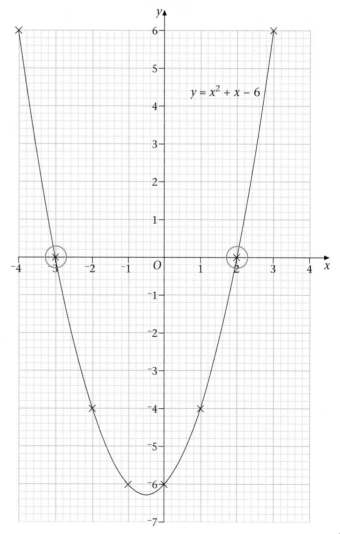

Practice question

1 For each of the following graphs:
   i   complete a table of coordinates for $x = -4$ to $x = +4$
   ii draw each graph on a new set of axes
   iii state the equation of the line of symmetry of each graph
   iv use the graph to find the solutions of $y = 0$.

   a $y = x^2 - 6$ **b** $y = x^2 - 2x - 3$ **c** $y = 8 - x^2$ **d** $y = 7 - \frac{1}{2}x^2$

Practice exam questions

1 **a** Copy and complete the table of values for $y = x^2 + 3$.

x	-2	-1	0	1	2	3
y	7	4				

   b Copy the grid, then plot the points from the table and draw the graph of $y = x^2 + 3$.

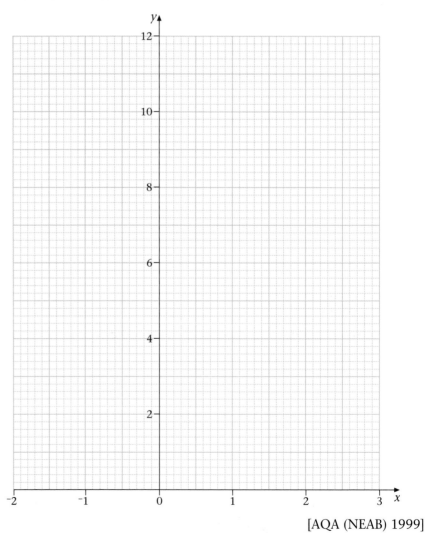

[AQA (NEAB) 1999]

2 a i Copy and complete the table for the graph of $y = x^2 - 7$.

x	-3	-2	-1	0	1	2	3
y	2	-3	-6				2

ii Copy the grid and draw the graph of $y = x^2 - 7$.

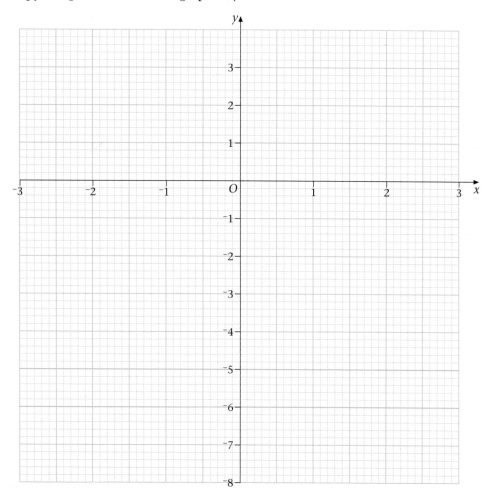

b Use your graph to solve the equation $x^2 - 7 = 0$.

c Use your graph to state the minimum value of y.

[AQA (NEAB) 2000]

3 **a** Copy and complete this table of values and use it to draw the graph of
$y = 5 + x - x^2$ on a copy of the grid below.

x	-3	-2	-1	0	1	2	3
y		-1	3	5	5		-1

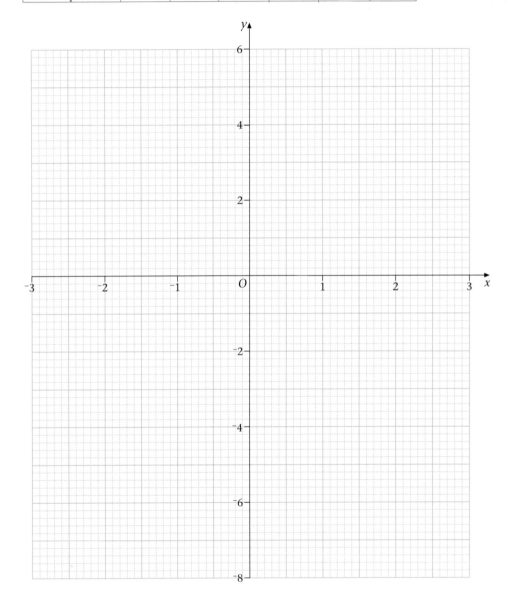

b Use your graph to find the values of x when $y = 0$. [AQA (SEG) 2000]

4 a Copy and complete this table of values for $y = x^2 - 2x - 5$.

x	-3	-2	-1	0	1	2	3	4
y		3	-2	-5		-5	-2	3

On a copy of the grid below, draw the graph of $y = x^2 - 2x - 5$.

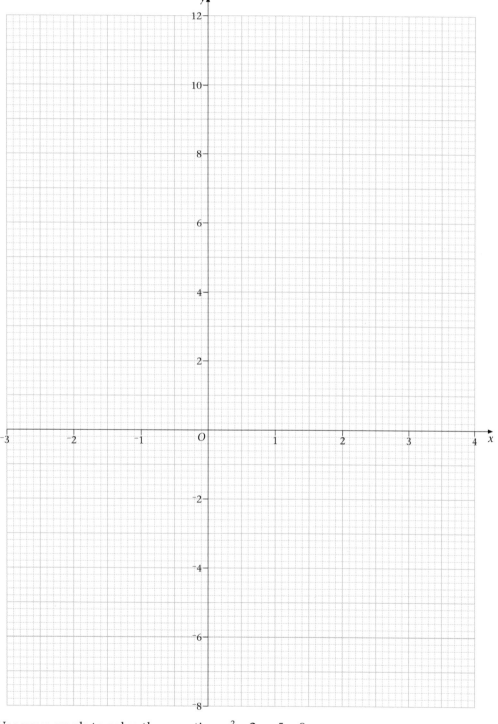

b Use your graph to solve the equation $x^2 - 2x - 5 = 0$.

[AQA (SEG) 2001]

5 This is a table of values for the function $y = 5 - \frac{1}{2}x^2$.

x	−4	−3	−2	−1	0	1	2
y	−3	0.5	3	4.5	5	4.5	3

  a On a copy of the grid below, plot the points shown in the table and hence draw the graph of $y = 5 - \frac{1}{2}x^2$ from $x = -4$ to $x = 2$.

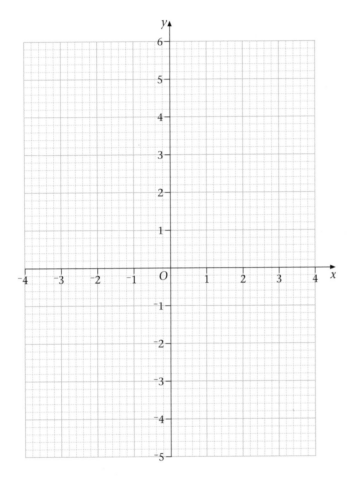

The graph is symmetrical about the y-axis.

  b Use this symmetry to plot the points where $x = 3$ and $x = 4$.

  c Read from your graph the solutions of the equation $5 - \frac{1}{2}x^2 = 0$.

[AQA (SEG) 2002]

22 Interpreting graphs

You may be required to plot graphs arising from real-life problems or you may be required to interpret graphs that model real-life situations. These graphs could be comparing:

- distance against time
- speed (velocity) against time
- different units (such as a conversion graph for converting £ to $)
- depth of water in a container against time
- different comparisons (such as height against age).

Distance–time graphs

Graphs can be plotted which show how the **distance** of an object from a fixed point varies with **time**. These are called **distance–time graphs**.

A horizontal line on a distance–time graph means that the object is stationary.

Example 22.1

The graph shows a cyclist's journey starting from a point *A* stopping at *B*, continuing to *C* and then returning to *A*.

a At what time did the journey start?

b How long was the stop at *B*?

c What was the total distance travelled?

d Calculate the speed from *A* to *B*.

e Which part of the journey was the fastest?

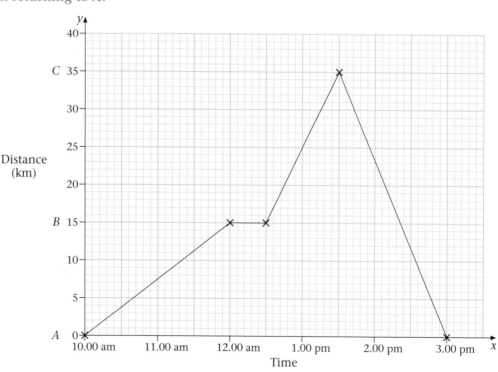

Solution

a 10.00 am, as this is the time of the first point on the graph.

b 30 minutes, as the graph is horizontal between 12.00 pm and 12.30 pm, which indicates that the cyclist is not moving.

c 70 km, as the cyclist travels 35 km from *A* to *C* and another 35 km on the return journey.

d The cyclist travels 15 km in two hours (10.00 pm to 12.00 pm), so the cyclist would have travelled $\frac{15}{2} = 7.5$ km in one hour, i.e. 7.5 km/h.

You can use the formula Speed $= \dfrac{\text{Distance}}{\text{Time}}$ to calculate this answer.

e The return journey from *C* to *A* was the fastest, as this is the steepest part of the graph.
Note: you do not have to calculate the speed of the cyclist for each section of the journey in order to answer this question, just compare the steepness of the lines.

Example 22.2

a A lorry travels 25 km in 25 minutes, stops for 10 minutes and then travels a further distance of 15 km in 20 minutes. Show this information on a grid.

b Calculate the lorry's average speed for the whole journey. Give your answer in km/h.

Solution

a The lorry travelled 25 km in 25 minutes so plot the point (25, 25) on the
graph. Join this point to the origin with a straight line.
The lorry stops for 10 minutes and so does not travel any more distance.
Draw a horizontal line from (25, 25) to (35, 25).
The lorry then travels a further distance of 15 km in another 20 minutes.
The lorry has now travelled a total distance of 25 + 15 = 40 km in a total
time of 25 + 10 + 20 = 55 minutes.
Plot the point (55, 40) and join this point to (35, 25) with a straight line.

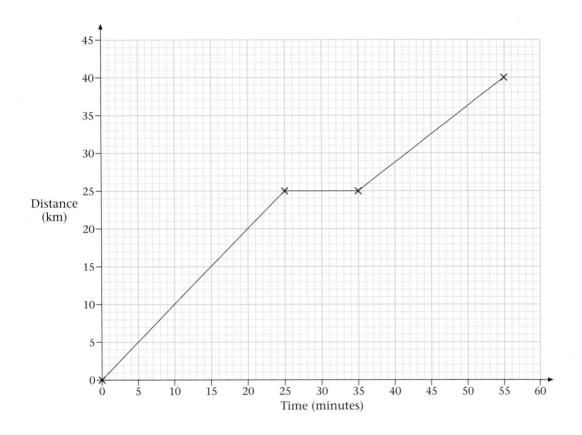

b Average speed $= \dfrac{\text{total distance}}{\text{total time}}$

$\qquad = \dfrac{40}{55}$

$\qquad = 0.7\dot{2} \text{ km/min}$

$\qquad = 0.7\dot{2} \times 60 \text{ km/h}$

$\qquad = 43.6 \text{ km/h}$

Example 22.3

A bus timetable is shown below.

Bus		Time	Distance travelled (km) between stops
Castleford	*depart*	0800	
Garforth	*arrive* *depart*	0815 0820	10 km
Leeds	*arrive* *depart*	0840 0850	8 km
Garforth	*arrive* *depart*	0910 0915	8 km
Castleford	*arrive*	0930	10 km

Draw an accurate distance–time graph showing the distance of the bus from Castleford for this bus journey.

Solution

You will normally be given the grid and the axes labelled in the exam.

You will always have time on the horizontal axis and distance on the vertical axis.

The time axis needs to start at 0800 and finish at 0930, which is 1.5 hours or 90 minutes.

Since the bus times are given in hours and minutes it may be better to label the time axis in minutes.

The distances are in kilometres.

The total journey from Castleford to Leeds is 18 km so the distance axis could be scaled up to 20 km.

The graph is shown opposite.

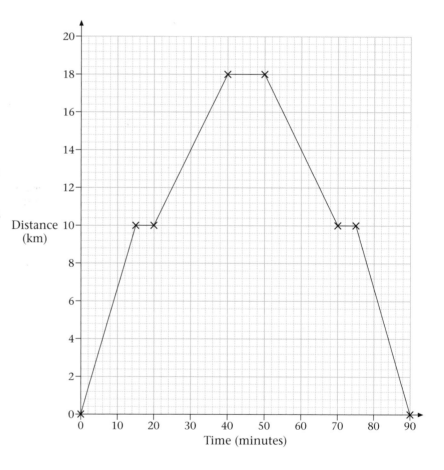

Practice questions 1

1 Andy leaves his home by car at 0800 and drives to work.
He travels 24 km in $\frac{3}{4}$ hour.
He stops because of an accident for 15 minutes and then continues his journey travelling 8 km in 20 minutes.

   a Draw a distance–time graph to show Andy's journey on a copy of the grid opposite.
   b Calculate Andy's average speed for the whole journey to work. Give your answer in km/h.

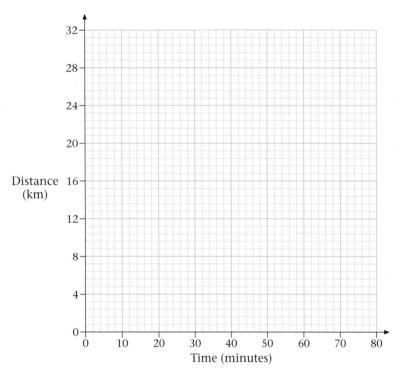

2 The graph shows the journey of a cyclist from Guildford to London.

   The distance from Guildford to London is 30 miles.

   a For how long did the cyclist stop during the journey?
   b What was the cyclist's average speed from *A* to *B*?
   c The cyclist stopped in London for 4 hours. He returned to Guildford at an average speed of 14 miles per hour. Calculate the time the cyclist arrived back in Guildford

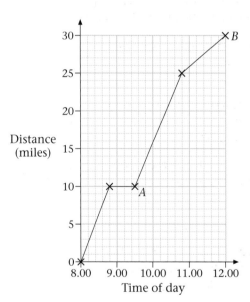

153

Speed–time graphs (or velocity–time graphs)

Sometimes you may be asked to interpret or draw a **speed–time graph**.

Speed will always be on the vertical axis and time will always be on the horizontal axis.

A horizontal line on a speed–time graph means that the speed is constant.

Example 22.3

A speed–time graph of a motorist travelling from London to Leicester is shown.

Interpret this graph and describe the motorist's speed during this journey.

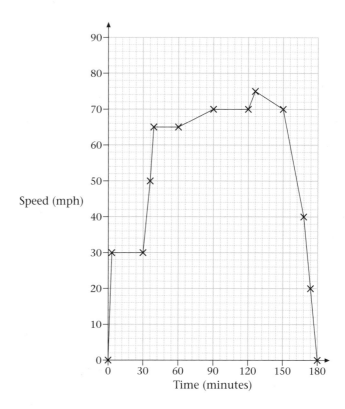

Solution

The motorist sets off from London gradually increasing her speed from 0 mph to 30 mph over a 3 minute period.

She then maintains a steady speed of 30 mph for 27 minutes.

Then she increases her speed gradually to 50 mph by 36 minutes.

Between 36 minutes and 39 minutes she accelerates from 50 mph to 65 mph.

She maintains this speed for 21 minutes.

From 1 hour to 1.5 hours she accelerates steadily to a speed of 70 mph, which she maintains for 30 minutes.

During the next 30 minutes she increases her speed to a maximum speed of 75 mph and then gradually slows down to 70 mph.

For the last 30 minutes of her journey she slows down steadily to arrive at Leicester after 3 hours driving.

Example 22.4

A cyclist starts from rest and accelerates steadily to a speed of 12 km/h within 5 minutes.

He cycles at 12 km/h for the next 20 minutes and then slows down steadily to a stop during the next 5 minutes.
After 15 minutes resting, the cyclist gradually increases his speed to 15 km/h over 10 minutes.
He maintains this speed for another 10 minutes and then steadily slows down to stop over the last 15 minutes of this journey.

Draw a speed–time graph to show this cyclist's journey.

Solution

Note that speed is on the vertical axis and must start from zero and go up to 15 km/h. Time is on the horizontal axis from zero up to
5 + 20 + 5 + 15 + 10 + 10 + 15 = 80 minutes.

Draw and label the axes and plot the relevant information in the correct positions on the grid.

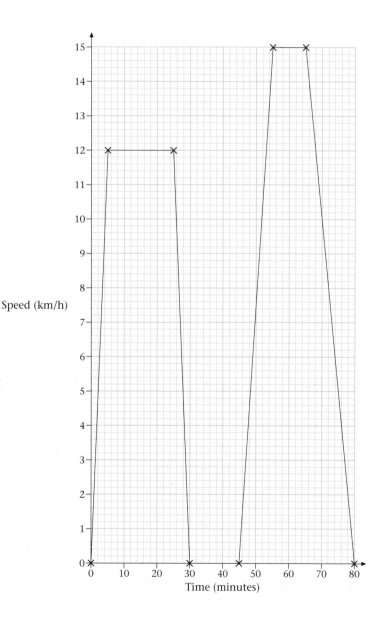

155

Practice questions 2

1   In the first lap of a race, a racing driver increases his speed from 0 mph to 100 mph in 8 seconds. He then increases his speed to a maximum speed of 150 mph in the next 20 seconds. He maintains this maximum speed for another 42 seconds. The racing driver then decreases his speed to zero over the next 30 seconds.

   Draw a speed–time graph to show the racing driver's speed during the first lap.

2   The speed–time graph shows the speed of a runner during the first hour of a 30 km marathon race. Describe the runner's speed during the first hour of the race.

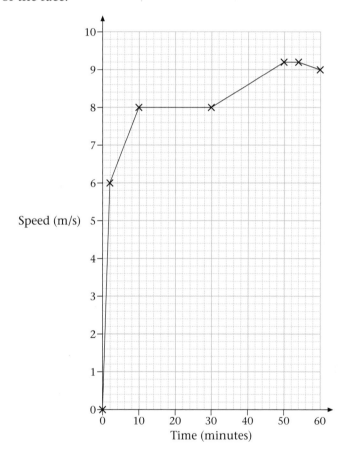

Conversion graphs

You can use straight line graphs (**conversion graphs**) to convert from one currency to another, e.g. from dollars to euros.

You can also use conversion graphs to change imperial units to metric units, e.g. from pounds to kilograms.

Reminder
Conversions were covered in Module 3.

Example 22.5

Steve is going to America for his holidays.

The exchange rate is £1 = $1.62.

a Draw a conversion graph of £ sterling to $.

b Use the graph to find:
   i how many dollars Steve would get for £500
   ii how many pounds would be equivalent to $350.

Solution

a The exchange rate is £1 = $1.62.
Calculate three points to draw the straight line graph.
Since £1 = $1.62
multiply both sides of this equation by 100 to get
    £100 = $162
multiply both sides of the original equation by 200 to get
    £200 = $324
multiply both sides of the original equation by 500 to get
    £500 = $810
These are the three points to plot on the axes.

b i Use the graph by locating £500 on the horizontal axis move up vertically to meet the graph and then read across to the vertical axis to find out how many dollars Steve would get for £500.
This was a point that was calculated to draw the graph.
   £500 = $810

   ii Locate $350 dollars on the vertical axis. It is equivalent to about £215.

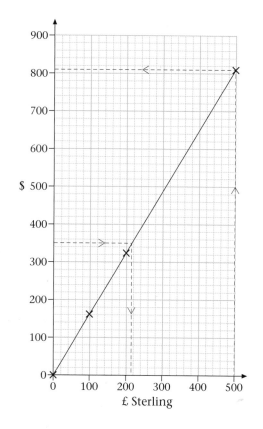

Practice question 3

1 Copy the graph and use it to answer these questions.

   a Convert the following temperatures to degrees Celsius.
   i 32° F
   ii 50° F
   iii 100° F
   iv 212° F

   b Convert the following temperatures to degrees Fahrenheit.
   i 20° C
   ii 30° C
   iii 5° C
   iv −10° C

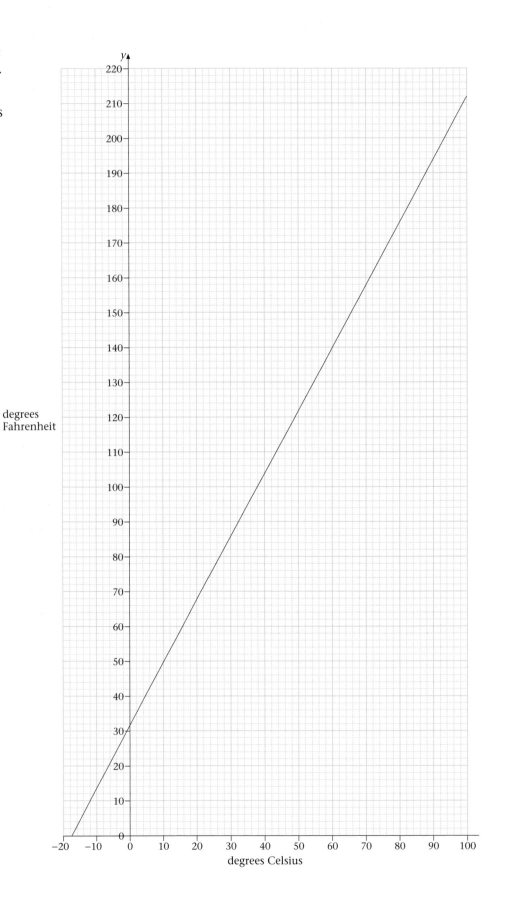

Drawing a graph of the depth of water in a container against time

If an empty container is being filled with water at a constant rate the depth of the water in the container at any time will depend upon the container's shape.

Example 22.6

Water is poured into a cylinder at a constant rate.
Sketch a graph of the depth of the water against time.

Solution

The cylinder has a uniform cross-section (a circle) and the water is poured into the cylinder at a constant rate. Therefore the depth of the water in the cylinder will also increase at a constant rate.

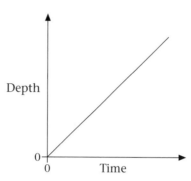

Example 22.7

A bird's water bowl has the shape shown opposite.
Water is poured into the empty bowl at a constant rate.
Sketch a graph of the depth of the water in the bowl against time.

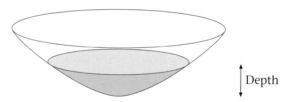

Solution

The bowl does not have a uniform cross-section. Since the water is poured in at a constant rate the depth of the water in the bowl will rise quickly and gradually slow down as the bowl is being filled.

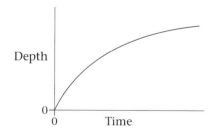

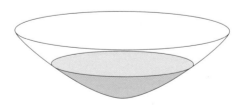

Practice questions 4

1   An empty wine bottle is being filled with wine at a constant rate.

Sketch a graph of the depth of the wine in the bottle against time.

2   Water is pouring out from a tap at the base of a cylindrical barrel.

Which graph, A, B or C, would show the depth of the water in the barrel against time?

Explain your answer.

A

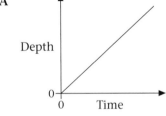

B

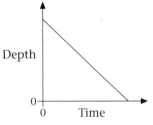

C

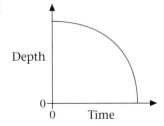

Different comparisons (such as height against age)

You may be asked to draw a sketch of a graph of, for example, a person's height against their age, or the height of a tree against its age.

Example 22.8

A boy's height is measured at various ages from 2 years old to 18 years old.
The pictogram illustrates the boy's height at certain ages.
Sketch a graph of the boy's height against his age.

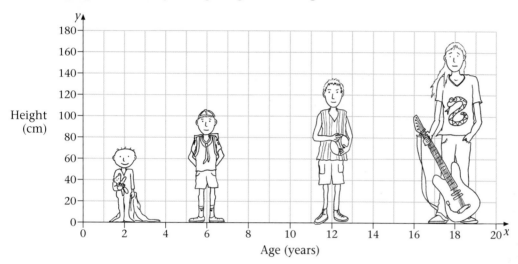

Solution

A baby boy at birth would probably be less than 40 cm tall, so mark a sensible point on the height axis when the boy's age is zero.

The height of each picture represents the boy's height at that age. Plot the points on a grid and join them with a smooth curve.

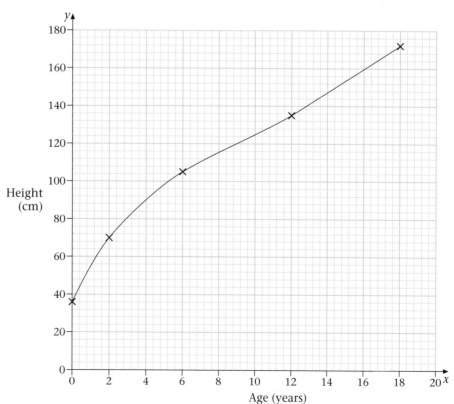

161

Practice question 5

1   The height of a tree after a
    certain number of years is
    shown in the pictogram.
    Sketch a line graph of the
    height of the tree against its
    age.

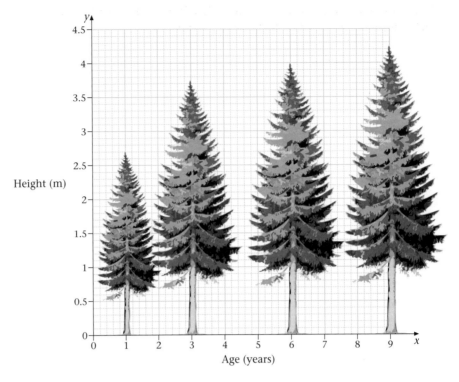

Height (m)

Age (years)

Practice exam questions

1   Errol leaves Fulling at 1000 to walk along the road to Graymer.
    The line on the graph represents his journey.

    a   At what time does Errol reach Graymer?
    b   How far is Errol from Fulling at 1037?
    c   Calculate Errol's speed in kilometres per hour.

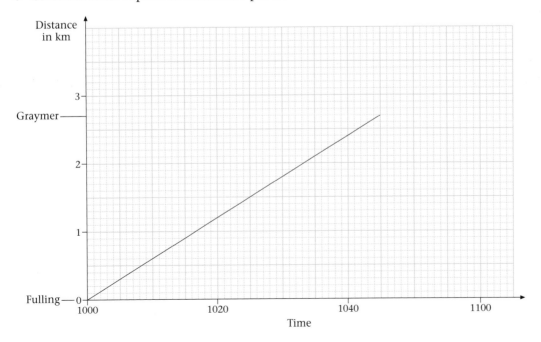

[AQA (SEG) 2001]

2 The graph shows the journey of a cyclist from Halton to Kendal.
The distance from Halton to Kendal is 30 miles.

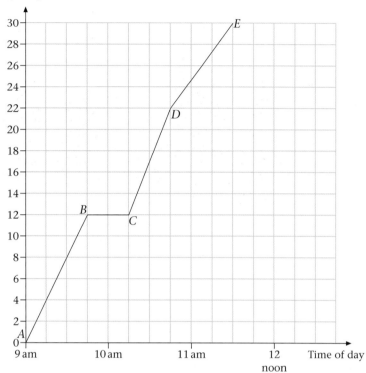

a For how long did the cyclist stop during the journey?

b What was the average speed for the part of the journey from *A* to *B*?

c On which section of the journey was the cyclist travelling at his fastest speed?
Explain clearly how you got your answer.

d The cyclist stayed in Kendal for 2 hours.
He then returned to Halton, without stopping, at an average speed of 12 miles per hour.
Calculate the time he arrived back in Halton.

[AQA (NEAB) 2001]

3 Jack and Jill left the Youth Hostel at 1000 to walk to Paildon.
They arrived at Paildon at 1108. They got back to the Youth Hostel at 1220.
The graph shows their journey.

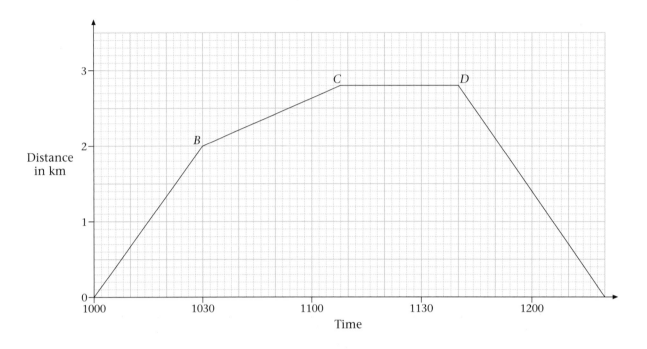

a How far is it from the Youth Hostel to Paildon?
b Describe *fully* what happened between points *C* and *D* on the graph.
c Describe what happened at point *B* on the graph.
d Calculate the speed, in kilometres per hour, at which they came back from Paildon to the Youth Hostel.

[AQA (SEG) 2000]

4   Kahlil cycles from Amwell to Barthorpe.

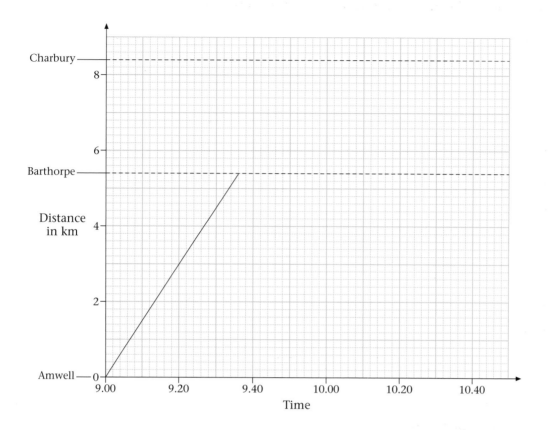

The graph shows the journey.
a How far is it from Amwell to Barthorpe?
b At what time does Kahlil arrive at Barthorpe?

Kahlil stays at Barthorpe for 20 minutes. Then he cycles on to Charbury at the same average speed as before.

c Show the rest of his journey on a copy of the grid.

[AQA (SEG) 2002]

5 Part of a train timetable is shown below.

Train		A	B	C	D
Pyeville	*depart*	0720	0735	0750	0805
Quiston	*arrive*	0733	↓	0803	0818
	depart			0807	0822
Robridge	*arrive*		↓	↓	0840
	depart				
Southerby	*arrive*		0825	0850	

a Which train, A, B, C or D, is shown in this distance–time graph?

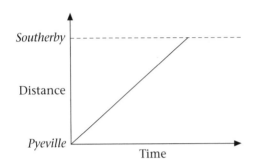

b Which train is shown in this distance–time graph?

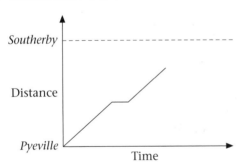

c The timetable for the next train is shown below.
Sketch the distance–time graph for train E on a copy of the axes below.

Train		E
Pyeville	*depart*	0820
Quiston	*arrive*	0833
	depart	0837
Robridge	*arrive*	0855
	depart	0900
Southerby	*arrive*	0932

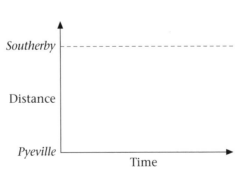

[AQA (SEG) 2001]

6 The graph illustrates a 1000 metre race between Nina and Polly.

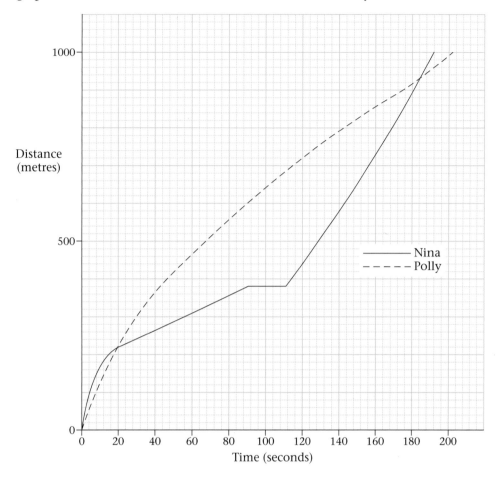

a Who was in the lead 10 seconds after the start of the race?
b Describe what happened 20 seconds after the start of the race.
c Describe what happened to Nina 90 seconds after the start of the race.
d Who won the race?

[AQA (NEAB) 1998]

7 A sky diver jumps from a plane.
The table shows the distance he falls, *d* metres, in *t* seconds.

t (seconds)	0	0.5	1.5	2.5	3.5
d (metres)	0	1	11	31	61

a Plot the points on a copy of the grid and join them with a smooth curve.
b Use your graph to find how many seconds he takes to fall 50 m.
c Use your graph to estimate how far he has fallen after 4 seconds.

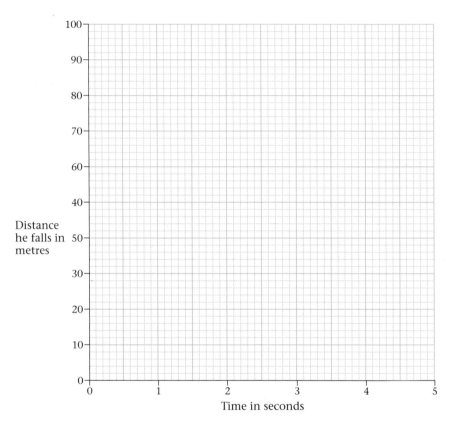

[AQA (NEAB) 1999]

8 A conversion graph is shown.

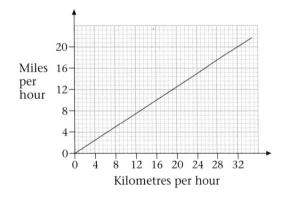

Use the graph to convert 96 kilometres per hour into miles per hour.

[AQA 2003]

23 Pythagoras' theorem

Pythagoras was a Greek mathematician.
He developed a mathematical theory about the lengths of the sides of
right-angled triangles.

A right-angled triangle has one angle of 90°.

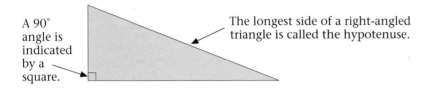

A 90°
angle is
indicated
by a
square.

The longest side of a right-angled
triangle is called the hypotenuse.

The hypotenuse (the longest side of a right-angled triangle) is the side that is
opposite the right angle.

Pythagoras discovered that if you square the length of the two short sides and
then add them together their sum will always be equal to the square of the
hypotenuse. This is called **Pythagoras' theorem**.

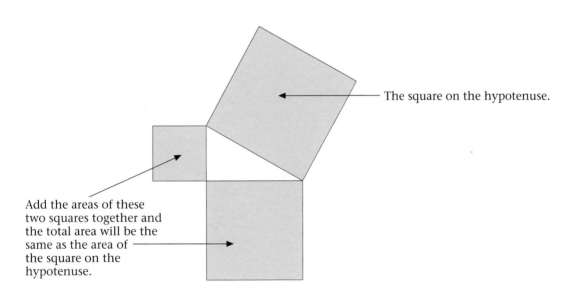

The square on the hypotenuse.

Add the areas of these
two squares together and
the total area will be the
same as the area of
the square on the
hypotenuse.

Pythagoras' general formula

Consider any right-angled triangle where the short sides have lengths a and b. The hypotenuse has length c.

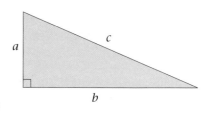

Then Pythagoras' formula is $c^2 = a^2 + b^2$.

This formula can be used whichever two sides are given, but c has to be the length of the hypotenuse.

Pythagoras' theorem enables you to calculate the length of a side of a right-angled triangle when you know the lengths of the other two sides.

Consider a right-angled triangle with the short sides of length 3 cm and 4 cm.

A drawing of the triangle with the squares on each side is shown below.

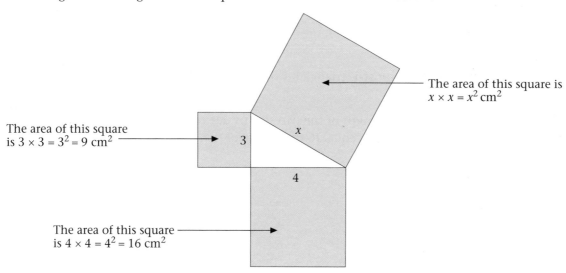

The area of this square is $x \times x = x^2$ cm^2

The area of this square is $3 \times 3 = 3^2 = 9$ cm^2

The area of this square is $4 \times 4 = 4^2 = 16$ cm^2

Pythagoras' theorem says that the sum of the squares of the two short sides is equal to the square of the hypotenuse.

Using the formula
for the above triangle you have

$$c^2 = a^2 + b^2$$
$$x^2 = 3^2 + 4^2$$
$$x^2 = 9 + 16$$
$$x^2 = 25$$

take the square root of both sides

$$x = \sqrt{25}$$
$$x = 5$$

So the length of the hypotenuse is 5 cm.

This particular right-angled triangle has sides of length 3 cm, 4 cm and 5 cm. This is a special right-angled triangle because it is the smallest right-angled triangle with integer lengths.

3, 4, 5 is called a Pythagorean triple. This triangle is sometimes referred to as a 3, 4, 5 triangle.

Questions on Pythagoras' theorem will appear on the non-calculator paper as well as the calculator paper.

The Pythagoras questions on the non-calculator paper will test that you know the squares of integers from 1 to 15 and that you know the square roots of the square numbers up to 225.

Example 23.1

Calculate the length of the missing side in the following triangle.

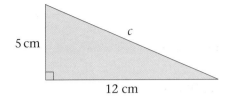

Solution

You have to recognise that the triangle is a right-angled triangle so that you can use Pythagoras' theorem.

Also notice that c is the hypotenuse.

Write down Pythagoras' formula. $c^2 = a^2 + b^2$
Substitute the lengths you know into the formula.

$$c^2 = 5^2 + 12^2$$
$$c^2 = 25 + 144$$
$$c^2 = 169$$
$$c = \sqrt{169}$$
$$c = 13 \text{ cm}$$

So the length of the hypotenuse is 13 cm.

EXAMINER **TIP**
5, 12, 13 is another well known Pythagorean triple.

Example 23.2

Calculate the length of the missing side in the following triangle.

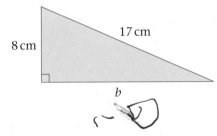

Solution

You have to recognise that the triangle is a right-angled triangle so that you can use Pythagoras' theorem.
You also have to realise that you are finding a short side.
Write down Pythagoras' formula.
Substitute the lengths you know into the formula.

$$c^2 = a^2 + b^2$$
$$17^2 = 8^2 + b^2$$
$$289 = 64 + b^2$$

Notice that this equation will need rearranging to find b.
Subtract 64 from both sides.

$$289 - 64 = b^2$$
$$225 = b^2$$
$$\sqrt{225} = b$$
$$15 = b$$

Reminder
Add the squares for the hypotenuse squared. Find the difference between the squares for a shorter length squared.

EXAMINER **TIP**
8, 15, 17 is another Pythagorean triple.

So here the length of the short side is 15 cm.

Practice question 1

1   Find the length of the missing side in each of the following triangles, without using a calculator.

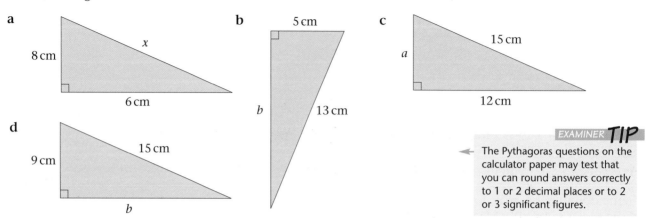

a

8 cm
x
6 cm

b

5 cm
b
13 cm

c

15 cm
a
12 cm

d

15 cm
9 cm
b

> EXAMINER **TIP**
> The Pythagoras questions on the calculator paper may test that you can round answers correctly to 1 or 2 decimal places or to 2 or 3 significant figures.

2   Find the length of the missing side in each of the following triangles.

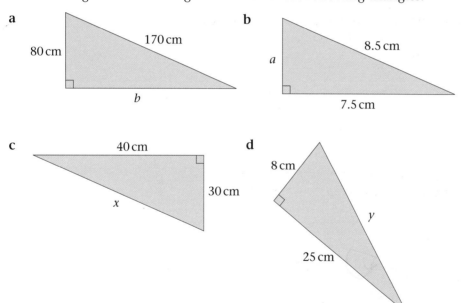

a

170 cm
80 cm
b

b

8.5 cm
a
7.5 cm

c

40 cm
x
30 cm

d

8 cm
y
25 cm

Checking if a triangle is a right-angled triangle

You can check if the lengths of the sides of a triangle will satisfy Pythagoras' theorem. If they do then the triangle must be a right-angled triangle.

Example 23.3

A triangle has sides 7 cm, 24 cm and 25 cm. Is the triangle a right-angled triangle? You must show your working.

Solution

It is useful to sketch the triangle and insert the lengths of the sides onto the sketch.
At this point the triangle looks as if it could be right-angled, but check with Pythagoras' theorem.

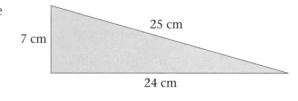

If the triangle is right-angled then $c^2 = a^2 + b^2$.

Calculate c and then check if it is equal to 25.

$c^2 = 7^2 + 24^2$
$c^2 = 49 + 576$
$c^2 = 625$
$c = \sqrt{625}$
$c = 25$ cm

Pythagoras' formula works and so this triangle must be a right-angled triangle and the right angle must be opposite the longest side.

Practice question 2

1 Use Pythagoras' theorem to find out which of the following triangles are right-angled, stating which angle is the right angle. You *must* show your working.

a

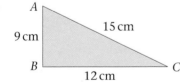

b

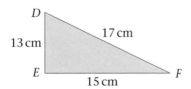

c

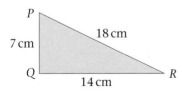

d

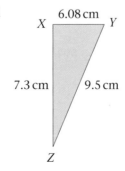

e

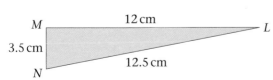

Finding the length of a line joining two points

When a line is drawn joining two sets of coordinates on a graph you can use Pythagoras' theorem to calculate the length of the line.

Example 23.4

Calculate the length of the line joining
$P(1, 3)$ to $Q(4, 6)$.

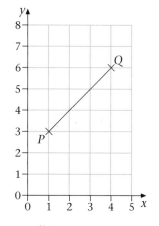

Solution

It is useful to draw a sketch of the line marking in the coordinates of the points P and Q.

Then draw in a horizontal line in the positive x-direction from P and a vertical line downwards from Q to form a right-angled triangle PQR.

Count the squares from P to R and from R to Q. These are the lengths of the two short sides of the right angled triangle.

The sides are both 3 units long.

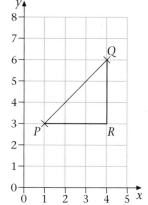

A better way to calculate the lengths of the two short sides is to find the difference between the x-coordinates and the y-coordinates of the points P and Q.

The coordinates of P are $(1, 3)$ and of Q are $(4, 6)$.

So the difference between the x-coordinates is
$4 - 1 = 3 = PR$

and the difference between the y-coordinates is
$6 - 3 = 3 = RQ$.

You can now use Pythagoras' theorem to work out the length of the line PQ.

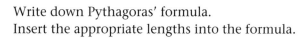

Write down Pythagoras' formula.
Insert the appropriate lengths into the formula.

$$c^2 = a^2 + b^2$$

$$PQ^2 = 3^2 + 3^2$$
$$PQ^2 = 9 + 9$$
$$PQ^2 = 18$$
$$PQ = \sqrt{18}$$
$$PQ = 4.2426...$$
$$PQ = 4.2 \text{ units (2 s.f.)}$$

So here the length of the line PQ is 4.2 units.

Example 23.5

The line $y = 8 - 2x$ cuts the y-axis at R and cuts the x-axis at S as shown on the grid below.

Calculate the length of the line segment RS.

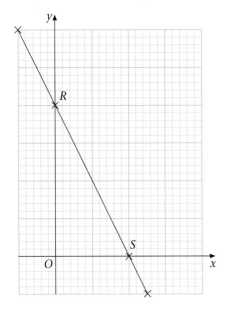

Solution

The equation of the line is $y = 8 - 2x$.

The intercept (where the line crosses the y-axis) is 8.

Where the line crosses the x-axis, $y = 0$

so $8 - 2x = 0$
$$8 = 2x$$
$$x = 4$$

Using the axes and the line segment RS draw a right-angled triangle.

Use Pythagoras' formula. $c^2 = a^2 + b^2$
Insert the appropriate lengths into the formula.
$$RS^2 = 8^2 + 4^2$$
$$RS^2 = 64 + 16$$
$$RS^2 = 80$$
$$RS = \sqrt{80}$$
$$RS = 8.944...$$
$$RS = 8.9 \text{ units (2 s.f.)}$$

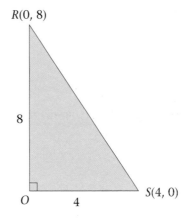

So here the length of the line segment RS is 8.9 units.

Practice questions 3

1   Calculate the length of the lines joining the following pairs of coordinates.

    a $A(1, 5)$ to $B(4, 1)$
    b $P(3, 4)$ to $Q(5, 7)$
    c $M(-4, -6)$ to $N(1, 6)$
    d $C(-2, 0)$ to $D(4, 0)$
    e $G(0, 0)$ to $H(5, 5)$

2   The line $y = 10 - 2x$ cuts the y-axis at P and cuts the x-axis at Q as shown on the grid below.

    Calculate the length of the line segment PQ.

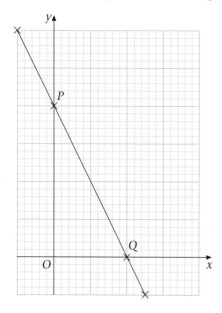

Practice exam questions

1   In a game two umpires stand on a pitch as shown.
    How far apart are the two umpires?

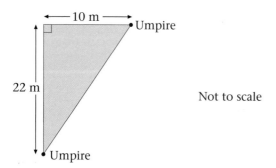

Not to scale

[AQA (NEAB) 1996]

2 The sketch shows the line $3x + 2y = 6$.
The line crosses the y-axis at A and the x-axis at B.

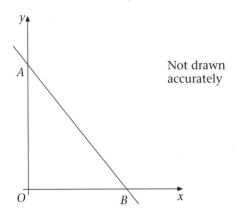

Not drawn
accurately

a Find the coordinates of A.   **b** Calculate the length of AB.   [AQA (NEAB) 2001]

3 PQR is the cross-section of a roof, with $PR = RQ$.
$PQ = 8.6$ m.
R is 3.6 m above the level of PQ.
Calculate the length of RQ.

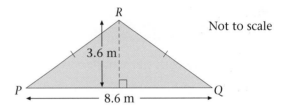

Not to scale

[AQA (SEG) 2002]

4 An oil rig is 15 kilometres East and 12 kilometres North from Kirrin.

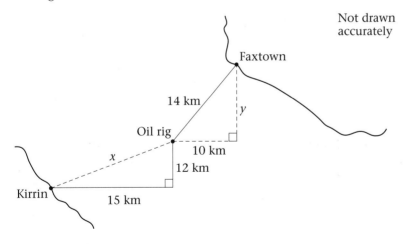

Not drawn
accurately

a Calculate the direct distance from Kirrin to the oil rig.
(The distance is marked x on the diagram.)
b An engineer flew 14 kilometres from Faxtown to the oil rig.
The oil rig is 10 kilometres West of Faxtown.
Calculate how far South the oil rig is from Faxtown.
(The distance is marked y on the diagram.)   [AQA (NEAB) 1999]

5 The sketch shows triangle *ABC*.
AB = 40 cm, *AC* = 41 cm and *CB* = 9 cm.

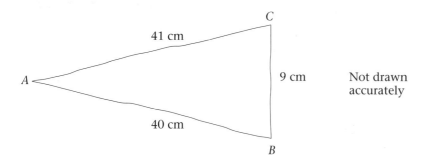

By calculation, show that triangle *ABC* is a right-angled triangle.

[AQA (NEAB) 2000]

24 Trigonometry

You use **trigonometry** with right-angled triangles.

You can find the length of a side or the size of an angle when given two pieces of information about the right-angled triangle.

Naming the sides and angles of a triangle

The sides of a right-angled triangle are given special names.
The longest side of a right-angled triangle is called the hypotenuse.

The two shorter sides are called the **opposite** side and the **adjacent** side depending upon which angle in the triangle you are using.

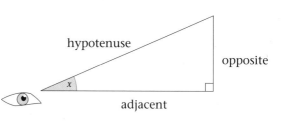

Imagine your eye is looking through the angle *x*. The side directly opposite angle *x* is called the opposite side.

The other short side is the adjacent side.

It is very important that you name the sides correctly.

EXAMINER **TIP**

← The word 'adjacent' means 'next to'. The adjacent side is the side 'next to' the angle you are using.

Example 24.1

You are using trigonometry in $\triangle ABC$ and are going to use angle A.

Name the sides of the triangle.

EXAMINER **TIP**

$\triangle ABC$ is short for triangle *ABC*.

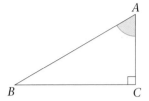

Solution

Imagine your eye is looking through the angle A. The side directly opposite angle A is called the opposite side. This is side BC.

The longest side is the hypotenuse which is always opposite the right angle. This is side AB.

It follows that the other short side is the adjacent side. This is side AC. It is the short side that is next to angle A.

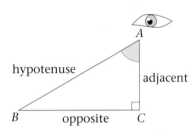

Practice question 1

1 Name all the sides in each triangle in relation to angle A.

a

b

c
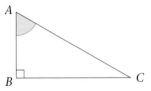

d

e
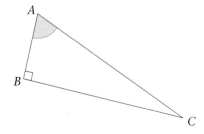

The three ratios

There are three ratios called **sine**, **cosine** and **tangent** that relate angles and sides in any right-angled triangle; these are defined below.

The sine ratio

For any right-angled triangle the **sine** ratio is defined as: $\dfrac{\text{opposite}}{\text{hypotenuse}}$.

This can be written as the formula $\sin A = \dfrac{\text{opposite}}{\text{hypotenuse}}$

where A is an acute angle in a right-angled triangle.

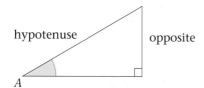

The **sine** ratio connects three pieces of information: the **angle A**, the length of the side **opposite** to angle A and the length of the **hypotenuse**.

If you know any two of these values, you can calculate the third value.

The cosine ratio

For any right-angled triangle the **cosine** ratio is defined as: $\dfrac{\text{adjacent}}{\text{hypotenuse}}$.

This can be written as the formula $\cos A = \dfrac{\text{adjacent}}{\text{hypotenuse}}$

where A is an acute angle in a right-angled triangle.

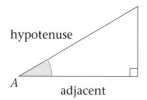

The **cosine** ratio connects three pieces of information: the **angle A**, the length of the side **adjacent** to angle A and the length of the **hypotenuse**.

If you know any two of these values you can calculate the third value.

The tangent ratio

For any right-angled triangle the **tangent** ratio is defined as: $\dfrac{\text{opposite}}{\text{adjacent}}$.

This can be written as the formula $\tan A = \dfrac{\text{opposite}}{\text{adjacent}}$

where A is an acute angle in a right-angled triangle.

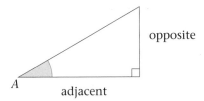

The **tangent** ratio connects three pieces of information: the **angle** A, the length of the side **opposite** to angle A and the length of the side **adjacent** to angle A.

If you know any two of these values, you can calculate the third value.

Remembering the ratios

In order to use trigonometry correctly you will need to remember which sides go with each ratio. Here are some ways of remembering the ratios.

- *Silly Old Hens Can't Always Have Their Own Acorns*
  Here the initial letter of each word stands for a word in the formulae, i.e.

  Silly Old Hens means Sine $= \dfrac{\text{Opposite}}{\text{Hypotenuse}}$

  Can't Always Have means Cosine $= \dfrac{\text{Adjacent}}{\text{Hypotenuse}}$

  Their Own Acorns means Tangent $= \dfrac{\text{Opposite}}{\text{Adjacent}}$

- *Six Old Horses Clumsy And Heavy Trod On Albert*
  Again, the first letter of each word stands for a word in the formula.

- *SOHCAHTOA*
  This is read by saying the groups of three letters together and so this tells you the formula for each ratio, i.e. 'SOH', 'CAH', 'TOA'.

Using the ratios to calculate the length of a side

To calculate the length of a side you need to know the size of one angle (not the right angle) and the length of one other side. The information you have been given in the question will tell you which ratio to use. You can use SOHCAHTOA to help you.

What you are given in the question	What you are asked to find	Ratio to use
	Find *b* opposite the angle	Sine (SOH)
	Find the hypotenuse	Sine (SOH)
	Find *a* adjacent to the angle	Cosine (CAH)
	Find the hypotenuse	Cosine (CAH)
	Find *a*	Tangent (TOA)
	Find *b*	Tangent (TOA)

Example 24.2

Calculate the length *x*.

Solution

In $\triangle ABC$ you are given two pieces of information:

hypotenuse $(AC) = 20$ cm

angle $A = 64°$

x is the *opposite* side to the given angle.

Using the sine ratio:

$$\sin A = \frac{\text{opposite}}{\text{hypotenuse}}$$

Substitute the given values.

$$\sin 64° = \frac{x}{20}$$

Rearrange by multiplying both sides by 20.

$$20 \times \sin 64° = x$$

Use your calculator to find $20 \times \sin 64°$.

$$17.98 \text{ cm} = x$$

Since the question gives the measurements to 2 significant figures it is sensible to give the answer to, say, 2 significant figures.

So $x = 18$ cm (to 2 s.f.)

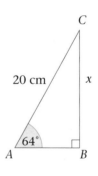

Reminder
Make sure you know how to use your calculator to get values of sine, cosine and tangents of angles.

Example 24.3

Calculate the length *y*.

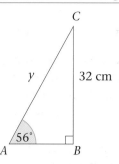

Solution

In △*ABC* you are given an angle and the length of the opposite side.

y is the hypotenuse.

Look at **SOH** CAH TOA. Since you have the **o**pposite side and you need to find the **h**ypotenuse you use the sine ratio.

$$\sin A = \frac{\text{opposite}}{\text{hypotenuse}}$$

Substitute the given values.

$$\sin 56° = \frac{32}{y}$$

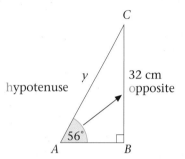

Rearrange by multiplying both sides by *y*.

$$y \times \sin 56° = 32$$

Rearrange by dividing both sides by sin 56°.

$$y = \frac{32}{\sin 56°}$$

Use your calculator to find $\frac{32}{\sin 56°}$.

y = 38.6 cm (3 s.f.)

> *EXAMINER* **TIP**
> Always check that the hypotenuse is the longest side.

Example 24.4

Calculate the length *x*.

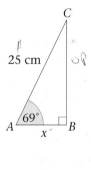

Solution

You are given the hypotenuse and the side *x* is adjacent to the angle. Look at SOH **CAH** TOA. Use the cosine ratio.

$$\cos A = \frac{\text{adjacent}}{\text{hypotenuse}}$$

$$\cos 69° = \frac{x}{25}$$

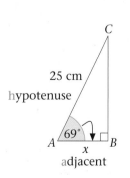

$$25 \times \cos 69° = x$$

8.96 cm = *x*

So *x* = 9.0 cm (to 2 s.f.)

> *EXAMINER* **TIP**
> Since the question gives the measurements to 2 significant figures it is sensible to give the answer to 2 significant figures.

Example 24.5

Calculate the length h.

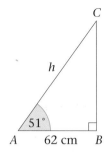

Solution

You are given an angle and the side adjacent to the angle.

h is the hypotenuse.

Look at SOH **CAH** TOA. Use the cosine ratio.

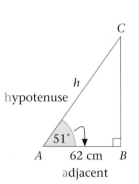

$$\cos A = \frac{\text{adjacent}}{\text{hypotenuse}}$$

$$\cos 51° = \frac{62}{h}$$

$$h \times \cos 51° = 62$$

$$h = \frac{62}{\cos 51°}$$

$$h = 98.5 \text{ cm (3 s.f.)}$$

Example 24.6

Calculate the length x.

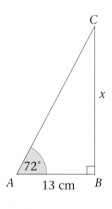

Solution

You are given an angle and the side adjacent to the angle.

x is the opposite side to the given angle.

Look at SOH CAH **TOA**. Use the tangent ratio.

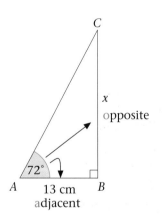

$$\tan A = \frac{\text{opposite}}{\text{adjacent}}$$

$$\tan 72° = \frac{x}{13}$$

$$13 \times \tan 72° = x$$

$$40.0 \text{ cm} = x$$

So $x = 40$ cm (to 2 s.f.)

EXAMINER **TIP**

Since the question gives the measurements to 2 significant figures it is sensible to give the answer to 2 significant figures.

Example 24.7

Calculate the length *v*.

Solution

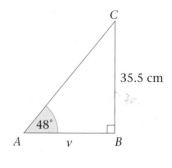

You are given an angle and the side opposite the angle.

v is the length of the adjacent side.

Look at SOH CAH TOA. Use the tangent ratio.

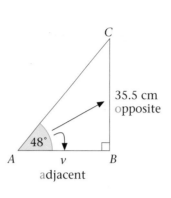

$$\tan A = \frac{\text{opposite}}{\text{adjacent}}$$

$$\tan 48° = \frac{35.5}{v}$$

$$v \times \tan 48° = 35.5$$

$$v = \frac{35.5}{\tan 48°}$$

$$v = 32.0 \text{ cm (3 s.f.)}$$

Example 24.8

In the diagram, $AB = 35$ cm, angle $BCD = 47°$

Angle $BAD = 38°$, BD is perpendicular to AC.

Calculate the length, *x*, of BC.

Give your answer to 1 decimal place.

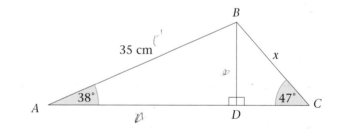

Solution

There is not enough known information about triangle BCD to obtain *x* directly. Use trigonometry with triangle ABD to find the length of BD.

$$\sin A = \frac{\text{opposite}}{\text{hypotenuse}}$$

$$\sin 38° = \frac{BD}{35}$$

You are given the angle and the hypotenuse, and if you need to find the opposite, BD.

$$BD = 35 \times \sin 38°$$

Look at SOH CAH TOA. Use the sine ratio.

$$BD = 21.548 \text{ cm}$$

Now that BD has been found, use it with triangle BCD to find *x*.

$$\sin A = \frac{\text{opposite}}{\text{hypotenuse}}$$

Sketching out triangle BCD will help you visualise the problem better.

$$\sin 47° = \frac{21.548}{x}$$

You know BD which is opposite the angle. *x* is the hypotenuse.

$$x = \frac{21.548}{\sin 47°}$$

Look at SOH CAH TOA. Use the sine ratio.

$$x = 29.5 \text{ cm (1 dp)}$$

So $x = 29.5$ cm (1 dp).

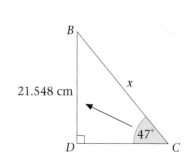

185

Practice questions 2

1 Calculate the length *z*.

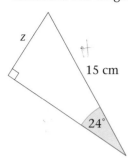

2 Calculate the length *a*.

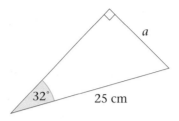

3 Calculate the length *p*.

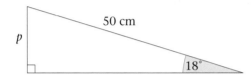

4 Calculate the length *t*.

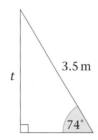

5 Calculate the length *b*.

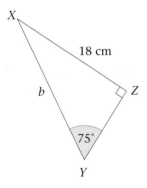

6 Calculate the length *g*.

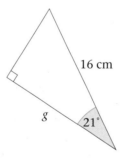

7 Calculate the length *j*.

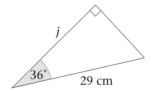

8 Calculate the length *k*.

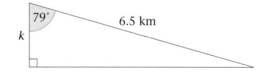

9 Calculate the length *r*.

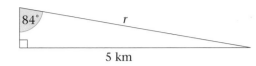

10 Calculate the length *s*.

11 Calculate the length *t*.

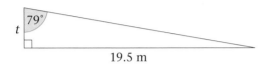

12 Calculate the length *s*.

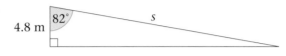

13 Find the length of the ladder.

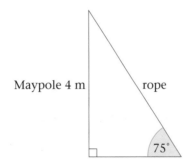

14 A rope is attached to the top of a maypole 4 m high. The rope makes an angle of 75° to the horizontal and is fixed to the ground. Calculate the length of the rope.

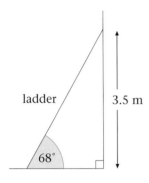

15 Find the length of the ladder.

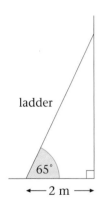

16 A rope is attached to the top of a ship's mast 7 m high and vertically above the deck of the ship. The rope makes an angle of 15° to the vertical mast and is fastened to the deck of the ship. Calculate the length of the rope from the mast to the deck.

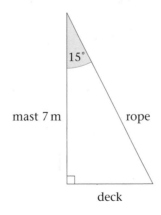

17 Find how far up the wall the ladder is touching.

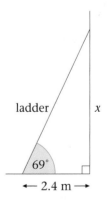

Using a ratio to calculate an angle

To calculate an angle using one of the ratios you must know the lengths of two sides. If you are asked to calculate an angle, you will be given the lengths of two sides. The sides you have been given will tell you which ratio to use.

What you are given	Ratio to use
One side *opposite* the angle and the *hypotenuse*	Sine
One side *adjacent* to the angle and the *hypotenuse*	Cosine
One side *adjacent* to the angle and one side *opposite* it	Tangent

> **Reminder**
> Use SOHCAHTOA to help you remember the ratios.

When you calculate an angle you will need to use the inverse of a ratio. For example, you may know for a triangle that:

$$\sin A = \frac{3}{5}$$

The angle A is found by applying the inverse sine function to each side.

$$A = \sin^{-1}\left(\frac{3}{5}\right)$$

There will be an inverse function for all ratios on your calculator. Often it will be accessed by pressing the shift or 2nd button followed by the ratio button, e.g. for $\sin^{-1}$, press [2nd] then [sin].

Example 24.9

Calculate angle A.

Solution

You are asked to calculate angle A. You are given the side opposite the angle and the hypotenuse.

Look at **SOH** CAH TOA. Use the sine ratio.

$$\sin A = \frac{\text{opposite}}{\text{hypotenuse}}$$

Substitute the given values.

$$\sin A = \frac{4}{5}$$

$$\sin A = 0.8$$

Now take the inverse sine of both sides.

$$A = \text{inv} \sin 0.8$$

Use your calculator to find inv sin 0.8.

inv sin 0.8 = 53.1° (1 d.p.)

So A = 53° (2 s.f.)

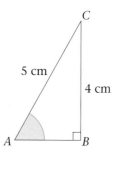

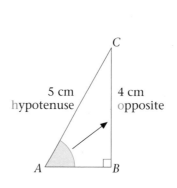

> **Reminder**
> $\sin^{-1}$ is often called inverse sin.

> EXAMINER *TIP*
> There will be an inverse sine button on your calculator: [sin⁻¹]. Make sure you know how to use it.

Example 24.10

Calculate angle A.

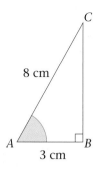

Solution

You are asked to calculate angle A. You are given the side adjacent to the angle and the hypotenuse.

Look at SOH **CAH** TOA. Use the cosine ratio.

$$\cos A = \frac{\text{adjacent}}{\text{hypotenuse}}$$

$$\cos A = \frac{3}{8}$$

$$\cos A = 0.375$$

$$A = \text{inv} \cos 0.375$$

$$A = 68° \text{ (2 s.f.)}$$

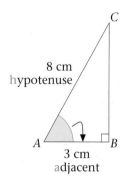

> **Reminder**
> $\cos^{-1}$ is often called inverse cos.

> *EXAMINER* **TIP**
> There will be an inverse cosine button on your calculator: $\boxed{\cos^{-1}}$. Make sure you know how to use it.

Example 24.11

Calculate angle A.

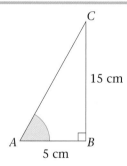

Solution

You are asked to calculate angle A. You are given the side opposite the angle and the side adjacent to the angle.

Look at SOH CAH **TOA**. Use the tangent ratio.

$$\tan A = \frac{\text{opposite}}{\text{adjacent}}$$

$$\tan A = \frac{15}{5}$$

$$A = \text{inv} \tan \frac{15}{5}$$

$$A = 72° \text{ (2 s.f.)}$$

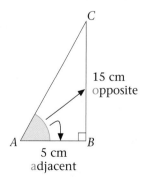

> **Reminder**
> $\tan^{-1}$ is often called inverse tan.

> *EXAMINER* **TIP**
> There will be an inverse tangent button on your calculator: $\boxed{\tan^{-1}}$. Make sure you know how to use it.

Practice questions 3

1   Calculate the angle *x* in each triangle.

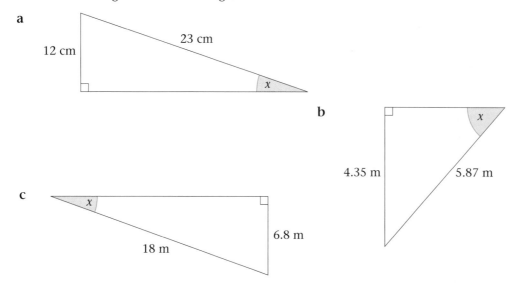

2   A ladder 3.45 m long is resting on horizontal ground and leaning against a vertical wall. The ladder reaches 3.2 m up the wall. What is the angle that the ladder makes with the ground?

3   Calculate the angle *x* in each triangle.

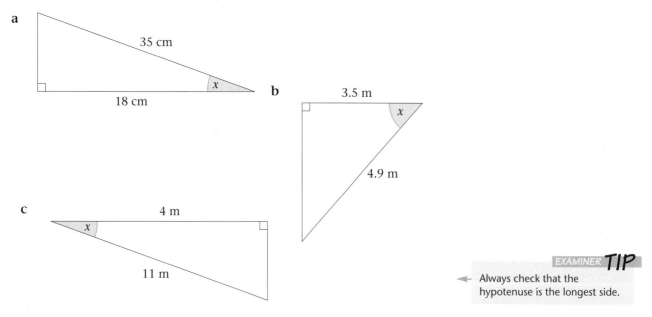

EXAMINER **TIP**

◄ Always check that the hypotenuse is the longest side.

4   A ladder 3.5 m long is resting on horizontal ground and leaning against a vertical wall. The ladder is 2.1 m away from the wall. What is the angle that the ladder makes with the ground?

5 Calculate the angle *x* in each triangle.

a

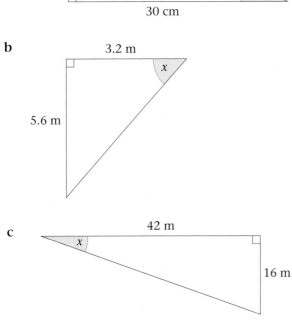

b

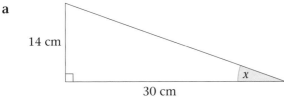

c

6 A builder needs to place his ladder 4.5 m up against a vertical wall. He places the base of the ladder 2 m away from the wall on horizontal ground. What is the angle that the ladder makes with the horizontal ground?

7 A man is standing on the top of a vertical cliff 50 m above sea level. A ship is 2.4 km from the base of the cliff. Calculate the angle of elevation of the ship to the man.

Practice exam questions

1 The diagram shows the side view of a wheelchair ramp.
The ramp makes an angle of 4° with the horizontal.
Calculate the length, marked *x*, of the ramp.
Give your answer in metres to a sensible degree of accuracy.

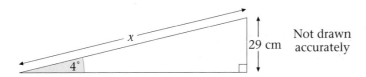

[AQA (NEAB) 2001]

2 Beverly is standing at *B*, 32 metres from the foot of a flagpole.
The flagpole is 7.5 metres high.

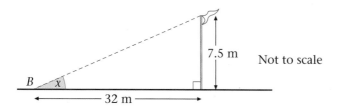

a Calculate *x*, the angle of elevation of the top of the flagpole from *B*.

A wire runs down to the ground at an angle of 47° to the flagpole as shown.

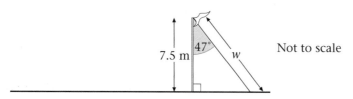

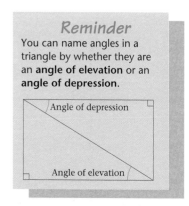

b Calculate *w*, the length of the wire.

[AQA (SEG) 2002]

3 The ladder leans against the side of a house.
The ladder is 4.5 m in length.

a The ladder makes an angle of 74° with the ground.
How high up the wall will it reach? (marked *x* in the diagram)
Give your answer to an appropriate degree of accuracy.

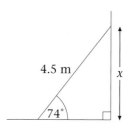

b The same ladder is now placed 0.9 m away from the side of the house.
What angle does the ladder now make with the ground? (marked *y* in the diagram)

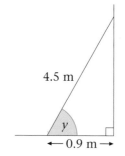

[AQA (NEAB) 2001]

4 a The triangle ABC is shown below.

    i Calculate the length of the side *BC*.
    ii Calculate the length of the side *AC*.

b The triangle *PQR* is shown below.

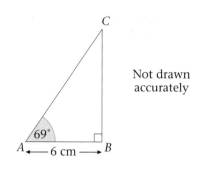

Not drawn accurately

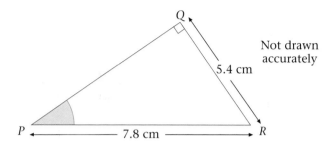

Not drawn accurately

    Calculate the size of angle *QPR*.

[AQA (NEAB) 1999]

5 The diagram shows part of a framework for a roof.
Triangles *ABC* and *CED* are right-angled.
AC = 3.2 m, *CE* = 4.1 m
Angle *ACB* is 49°. Angle *EDC* is 58°.

    a Calculate the length *BC*.
    b Calculate the length *CD*.

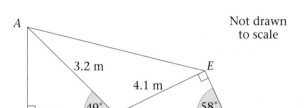

Not drawn to scale

[AQA (NEAB) 1998]

6 *ABD* and *BCD* are two right-angled triangles.
AB = 12 cm, *CD* = 8 cm, angle *BAD* = 30°
The two triangles are joined together as shown in the diagram.
ADC is a straight line.
Calculate the length *BC*, marked *x* on the diagram.

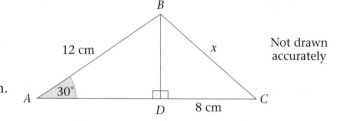

Not drawn accurately

[AQA (NEAB) 2002]

7 *A*, *B* and *C* are villages on the shores of a lake.
A is 9 km north of *B*.
C is 40 km due east of *B*.
    a Calculate the size of the angle *x*.
    b A boat sails directly from *A* to *C*.
        On its journey it passes *D*, the nearest point to *B*.
        Calculate the distance *BD*.

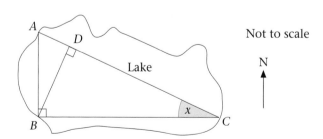

Not to scale

[AQA (NEAB) 1996]

25 Linear inequalities

An **inequality** is a mathematical statement using one of the following symbols.

Less than $<$

Less than or equal to $\leqslant$

Greater than $>$

Greater than or equal to $\geqslant$

To **solve** an inequality the method is the same as for solving equations except that it is not valid to multiply or divide through the inequality by a negative number. Multiplying or dividing by a negative number changes the inequality sign, e.g. if you divide both sides of $3 < 5$ by -1 it becomes $-3 > -5$.

Solutions can be written either using algebra or by showing the solution on a number line. For example:

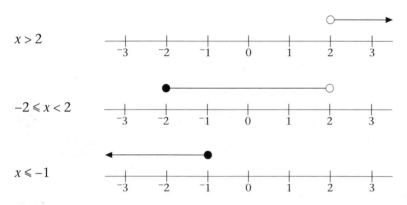

$x > 2$

$-2 \leqslant x < 2$

$x \leqslant -1$

An open circle ($\bigcirc$) is used for less than ($<$) or greater than ($>$) and a closed circle ($\bullet$) for less than or equal to ($\leqslant$) or greater than or equal to ($\geqslant$).

Example 25.1

Solve the inequality $2x + 1 < 9$.

Solution

	Equation	Inequality
	$2x + 1 = 9$	$2x + 1 < 9$
Subtract 1 from both sides	$2x = 8$	$2x < 8$
Divide both sides by 2	$x = 4$	$x < 4$

EXAMINER *TIP*

In an inequality always keep the inequality sign throughout the working. Do not change it to an equals sign!

The solution to the equation $2x + 1 = 9$ is $x = 4$.

The solution to the inequality $2x + 1 < 9$ is $x < 4$.

Example 25.2

Solve the inequality $4x - 3 \geqslant 17$.

Solution

	Equation	Inequality
	$4x - 3 = 17$	$4x - 3 \geqslant 17$
Add 3 to both sides	$4x = 20$	$4x \geqslant 20$
Divide both sides by 4	$x = 5$	$x \geqslant 5$

The solution to $4x - 3 \geqslant 17$ is $x \geqslant 5$.

Example 25.3

Solve the inequality $3x - 4 > 5(x + 8)$.

Solution

	Equation	Inequality
	$3x - 4 = 5(x + 8)$	$3x - 4 > 5(x + 8)$
Remove the brackets	$3x - 4 = 5x + 40$	$3x - 4 > 5x + 40$
Subtract $3x$ from both sides	$-4 = 2x + 40$	$-4 > 2x + 40$
Subtract 40 from both sides	$-44 = 2x$	$-44 > 2x$
Divide both sides by 2	$-22 = x$	$-22 > x$

> **EXAMINER TIP**
> It is easier to collect the x-terms on the side that makes it a positive term.

The solution to $3x - 4 > 5(x + 8)$ is $-22 > x$ or $x < -22$.

> **EXAMINER TIP**
> $-22 > x$ means -22 is greater than x.

Example 25.4

Solve the inequality $2 - x > 6$.

Solution

	Equation	Inequality
	$2 - x = 6$	$2 - x > 6$
Add x to both sides	$2 = 6 + x$	$2 > 6 + x$
Subtract 6 from both sides	$-4 = x$	$-4 > x$

The solution to $2 - x > 6$ is $-4 > x$ or $x < -4$.

It is important when solving inequalities to make sure that the variable is taken to the side where it has a positive coefficient.

Example 25.5

List the integer values of n such that $-3 \leqslant 3n < 12$.

Solution

$$-3 \leqslant 3n < 12$$

Dividing throughout by 3: $\quad -1 \leqslant n < 4$

The integer values of n are $-1, 0, 1, 2, 3$.

> EXAMINER *TIP*
>
> $n < 4$ means that n is strictly less than 4 and cannot equal 4.

> *Reminder*
>
> Integers are whole numbers.

Example 25.6

List the values of n, where n is an integer, such that $5 < 3n + 2 \leqslant 11$.

Solution

$$5 < 3n + 2 \leqslant 11$$

Subtracting 2 throughout: $\quad 3 < 3n \leqslant 9$
Dividing throughout by 3: $\quad 1 < n \leqslant 3$

The integer values of n are 2 and 3.

Practice questions

1 Represent the following inequalities on a number line.

  a $x > 3$ **b** $x \leqslant 2$ **c** $-2 < x \leqslant 1$

2 List the integer values of n such that:

  a $-3 < n \leqslant 1$ **b** $2 \leqslant n < 6$ **c** $-2 < n < 2$

3 Solve the inequalities.

  a $3x - 2 > 7$ **b** $2n + 5 \leqslant 17$ **c** $2a + 5 \leqslant 14 - a$

4 Solve the inequalities.

  a $4 < 2x - 7 \leqslant 10$ **b** $-5 \leqslant 3x + 1 < 13$ **c** $-11 \leqslant 4x - 1 < 15$

5 List the value of n, where n is an integer, such that:

  a $1 < n - 1 \leqslant 3$ **b** $3 \leqslant 2n + 1 < 7$ **c** $-6 \leqslant 3n < 21$

Practice exam questions

1 List the integer values of n such that $-4 \leqslant 2n < 6$. [AQA (NEAB) 2002]

2 List the values of n, where n is an integer, such that $3 \leqslant n + 4 < 6$. [AQA (SEG) 1998]

3 List the values of *n*, where *n* is an integer, such that $1 \leqslant 2n - 3 < 5$.   [AQA (SEG) 1999]

4 Solve the inequality $3 < 2x + 1 < 5$.   [AQA (SEG) 2000]

5 Solve the inequality $x + 20 < 12 - 3x$.   [AQA (NEAB) 2000]

26 Graphs of linear inequalities

You will need to know how to represent inequalities on a graph.

A straight line drawn on a graph divides the graph into three sets of points.

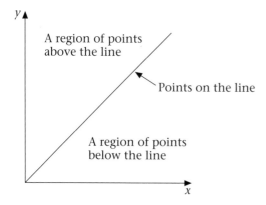

The points on the line are described by an equation, e.g. $y = x$. The regions above and below the line are described by inequalities.

The diagrams below show how to represent the different possible regions using the graph of $y = x$. Note that when the line does not form part of the region (when using < or >), you draw a dotted line. When the line does form part of the region (when using $\leqslant$ or $\geqslant$), you draw a solid line.

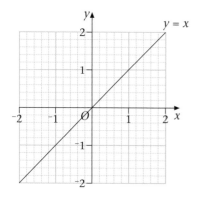

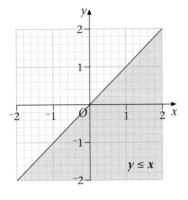

$y \leqslant x$

The line is a solid line so the shaded region includes points on the line.

At every point in the region the *y*-coordinate is less than or equal to the *x*-coordinate.

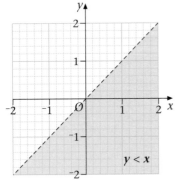

$y < x$

In this case, the line is dotted so the shaded region does not include points on the line.

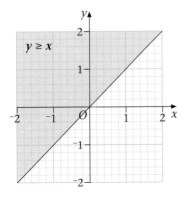

$y \geqslant x$

The line is a solid line so the region includes points on the line.

At every point in the region the *y*-coordinate is greater than or equal to the *x*-coordinate.

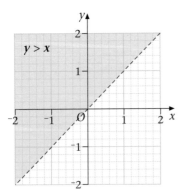

$y > x$

In this case the line is dotted so the shaded region does not include points on the line.

Example 26.1

On the grid, show the single region that is satisfied by all of these inequalities.

$x < 2$ $\qquad\qquad$ $y \geqslant -1$ $\qquad\qquad$ $y \leqslant x + 2$

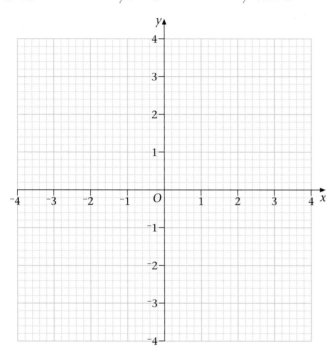

Solution

If you consider each inequality on its own, the separate graphs would look like this:

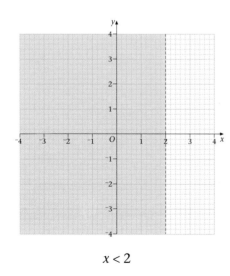

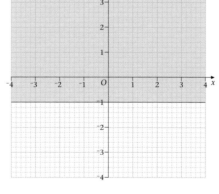

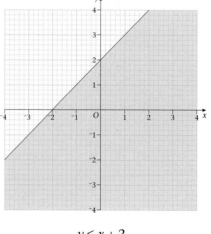

$\qquad\qquad x < 2 \qquad\qquad\qquad\qquad y \geqslant -1 \qquad\qquad\qquad\qquad y \leqslant x + 2$

Combining all three graphs onto one grid produces the following:
The shaded region is the single region satisfied by all three inequalities.

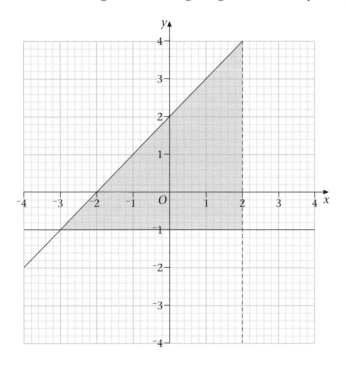

Example 26.2

Write down *four* inequalities which together describe the shaded area.

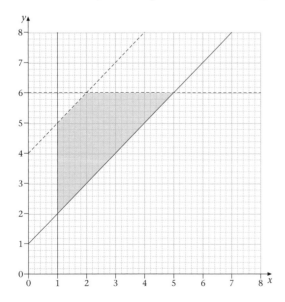

Solution

Labelling the four straight lines on the graph gives you the graph opposite.

The shaded region is to the *right* of the *solid line* $x = 1$, so you have $x \geqslant 1$.

The shaded region is *below* the *dotted line* $y = 6$, so you have $y < 6$.

The shaded region is *below* the *dotted line* $y = x + 4$, so you have $y < x + 4$.

The shaded region is *above* the *solid line* $y = x + 1$, so you have $y \geqslant x + 1$.

So the shaded region, bounded by these lines, can be described as:

$x \geqslant 1$ $y < 6$ $y < x + 4$ $y \geqslant x + 1$

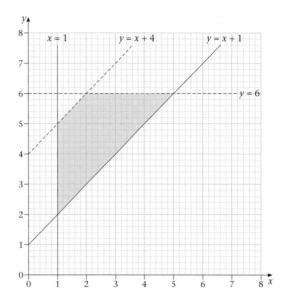

Practice questions

1 Draw graphs to show the region that is satisfied by each of the following inequalities.

    a $x \leqslant 2$ **b** $y < 1$ **c** $y > -x$ **d** $y \geqslant 2x + 1$

2 Draw graphs to show the single region that is satisfied by each of the following sets of inequalities.

    a $x \leqslant 4$ $y \leqslant 3$ $x + y \geqslant 1$
    b $x < 3$ $y \geqslant -2$ $y < x$
    c $2x + 3y < 6$ $x \geqslant -3$ $y \geqslant -1$
    d $y < 2x$ $y \geqslant x$ $y < 4 - 2x$

3 Write down the set of inequalities which together describe each shaded area.

a

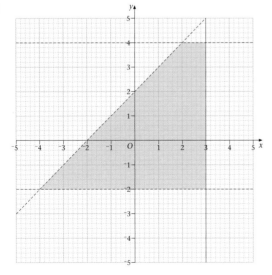

b

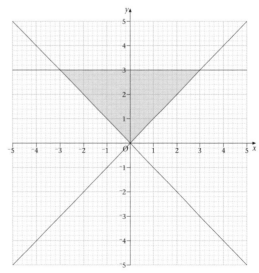

c

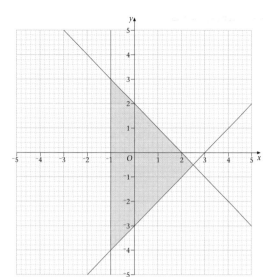

d

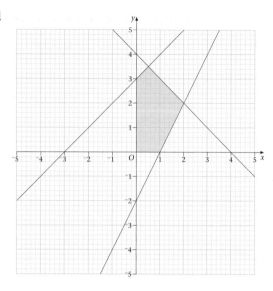

Practice exam questions

1 On a copy of each grid shade the region where:

a $x \geqslant 2$ b $y \leqslant 4$

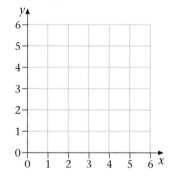 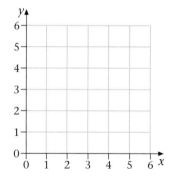

c $x + y \leqslant 4$

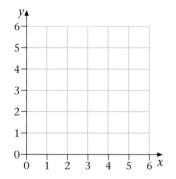

[AQA (NEAB) 1999]

2 **a** On a copy of the diagram draw and label the lines $x = 1$ and $x + y = 4$.

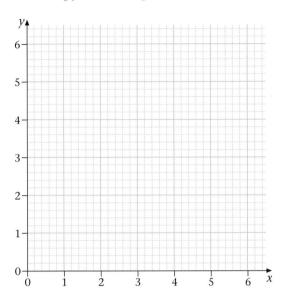

b Show clearly on the diagram the single region that is satisfied by all of these inequalities.

$y \geqslant 0$ $x \geqslant 1$ $x + y \leqslant 4$

Label this region R.   [AQA (SEG) 1999]

3 **a** List the integer values of n such that $3 \leqslant 3n < 18$.

b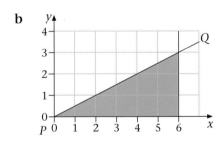

  i Find the equation of the line PQ.

  ii Write down *three* inequalities which together describe the shaded area.   [AQA 2003]

27 Trial and improvement

Trial and improvement is a method of finding a solution to an equation by substituting trial values into one side of the equation to try to make the equation equal to the value on the other side of the equation.

Questions will normally be set out using a table as shown below. You will be allowed to use your calculator.

Solve using trial and improvement the equation $x^2 + x = 17$, giving your answer to 1 decimal place.

The questions will always state the level of accuracy required.

Sometimes you will be given the first line to guide you.

x	Left side of equation $x^2 + x$	Comment
3	12	too small

Before you start to improve on the attempt $x = 3$, it is worth checking how the answer 12 was obtained to make sure you are substituting correctly before moving on to the rest of the table.

Replacing x with 3 in the left side of the equation gives

$3^2 + 3 = 9 + 3$
$\quad\ = 12$

12 is too small, since you are looking for an answer of 17.

Now try $x = 4$. Completing the next line gives:

x	Left side of equation $x^2 + x$	Comment
3	12	too small
4	$4^2 + 4 = 20$	too big

12 is too small and 20 is too big.

This now means that you know the answer is between $x = 3$ and $x = 4$.

Now improve the trial by trying $x = 3.5$.

x	Left side of equation $x^2 + x$	Comment
3	12	too small
4	$4^2 + 4 = 20$	too big
3.5	$3.5^2 + 3.5 = 15.75$	too small

EXAMINER TIP
The next value does not have to be 3.5; if you think the answer is more or less than this try a different value between 3 and 4.

15.75 is too small and 20 is too big.

This now means that the answer is between $x = 3.5$ and $x = 4$.

Improve the trial by trying $x = 3.6$.

x	Left side of equation $x^2 + x$	Comment
3	12	too small
4	$4^2 + 4 = 20$	too big
3.5	$3.5^2 + 3.5 = 15.75$	too small
3.6	$3.6^2 + 3.6 = 16.56$	too small

16.56 is too small and 20 is too big.
This now means that the answer is between $x = 3.6$ and $x = 4$.

Improve the trial by trying $x = 3.7$.

x	Left side of equation $x^2 + x$	Comment
3	12	too small
4	$4^2 + 4 = 20$	too big
3.5	$3.5^2 + 3.5 = 15.75$	too small
3.6	$3.6^2 + 3.6 = 16.56$	too small
3.7	$3.7^2 + 3.7 = 17.39$	too big

EXAMINER TIP
Having reached this stage you will have been awarded 2 marks.

16.56 is too small and 17.39 is too big.

This now means that the answer is between $x = 3.6$ and $x = 3.7$.

You now need to decide whether the answer to 1 decimal place is $x = 3.6$ or $x = 3.7$.

It is essential that you now try the mid-value of these, $x = 3.65$. If you do not do this you will lose the final mark as it is not enough to say that it is nearer to $x = 3.7$ even if this turns out to be the correct answer.

x	Left side of equation $x^2 + x$	Comment
3	12	too small
4	$4^2 + 4 = 20$	too big
3.5	$3.5^2 + 3.5 = 15.75$	too small
3.6	$3.6^2 + 3.6 = 16.56$	too small
3.7	$3.7^2 + 3.7 = 17.39$	too big
3.65	$3.65^2 + 3.65 = 16.9725$	too small

16.9725 is too small and 17.39 is too big.

This now means that the answer is between $x = 3.65$ and $x = 3.7$.

If you round any number between 3.65 and 3.7 to 1 decimal place the answer is 3.7.

So the solution to the equation is $x = 3.7$ to 1 decimal place.

Example 27.1

Solve $x^2 = 54$, giving your answer to 1 decimal place.

Solution 1

x	Left side of equation x^2	Comment
7	$7^2 = 49$	too small
8	$8^2 = 64$	too big
7.5	$7.5^2 = 56.25$	too big
7.4	$7.4^2 = 54.76$	too big
7.3	$7.3^2 = 53.29$	too small
7.35	$7.35^2 = 54.0225$	too big

As $x = 7.35$ is too big and $x = 7.3$ is too small, this means the answer is between $x = 7.3$ and $x = 7.35$.

So the solution to the equation is $x = 7.3$ to 1 decimal place.

Practice questions

Solve the following equations using trial and improvement. In each case use a table and give your answer to 1 decimal place.

1 $x^2 = 14$ **2** $x^3 + x = 20$ **3** $x^3 - 2x = 12$ **4** $x^3 + x^2 = 50$

Practice exam questions

1 Aisha is using trial and improvement to find a solution to the equation

$x^3 - 2x = 14$

This table shows her first two tries.
Continue the table to find a solution to the equation.

x	$x^3 - 2x$	Comment
2	4	too small
3	21	too big

Give your answer correct to 1 decimal place.   [AQA (NEAB) 2001]

2 Use trial and improvement to find a solution to the equation

$x^3 - 3x = 9$

Give your answer correct to 1 decimal place.

x	$x^3 - 3x$	Comment
2	2	too low
3	18	too high

[AQA (NEAB) 2001]

3 Stephanie is using trial and improvement to find a solution to the equation

$x^3 + 2x - 7 = 0$

The table shows her first two tries.
Continue the table to find a solution to the equation.

x	$x^3 + 2x - 7$	Comment
1	−4	too small
2	5	too big

Give your answer correct to 1 decimal place.   [AQA (NEAB) 2002]

28 Constructions

A **construction** question will usually ask you to 'use ruler and compasses only'. This is because these questions are testing whether you know how to construct certain angles *without* using a protractor.

There are several standard constructions involving angles of 90° and 60° which you should learn but you may also be asked to construct a shape such as a triangle.

If a question does not insist that you 'use ruler and compasses only' then you are allowed to use a protractor.

Many of these constructions work on the principle that you can draw points that are **equidistant** from another point using compasses. Equidistant means 'exactly the same distance'. For example, draw a line and mark two points on it. Now set your compasses and draw a circle around each point. Draw the circles large enough so that they overlap. The two points where the circles overlap are equidistant from the points on the line.

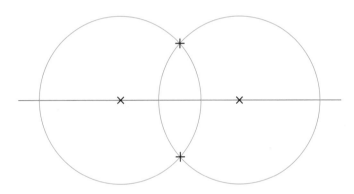

You will also come across two other terms; these are **angle bisectors** and **perpendicular bisector**.

An angle bisector is a line that cuts an angle exactly in two.

A perpendicular bisector cuts a line of a set distance exactly in half and at an angle of 90°.

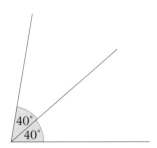

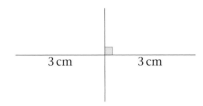

Standard constructions – step by step

60° at the end of a line

Keep the compass setting the same for Steps 1 to 4.

Step 1
The line will already be drawn for you.

_____

Step 2
Put your compass point on the end of the line where you want to form the angle and draw an arc. Make sure that your arc cuts the line as shown.

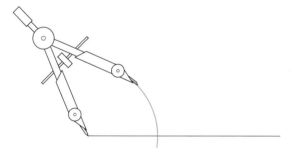

Step 3
Now put the compass point at the point where your arc cuts the line and draw another arc of equal radius. Make sure that the arcs intersect.

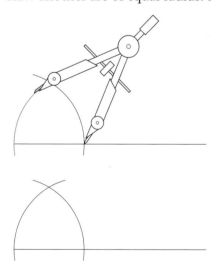

Step 4
The three points – two on the line and one where the two arcs intersect – make up three vertices of an equilateral triangle. Therefore if you join the point of intersection with one point on the line, you will make an angle of 60° with the line.

Use a ruler to complete the 60° construction.

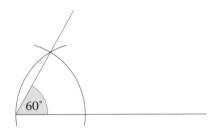

EXAMINER **TIP**

Remember you must use ruler and compasses. No arcs means no marks.

Equilateral triangle

Keep the compass setting the same throughout.

Step 1
Draw the base of the triangle *AB* 8 cm long.

Step 2
Open the compasses to a radius of 8 cm and put the point on *A*. Swing an arc from *B* above the base line.

Step 3
With the compasses at the same radius (8 cm) and the point on *B*, swing an arc from *A* to cut the first arc.

Step 4
Where the arcs cross is the third vertex, *C*, of the triangle. Join *C* to the other vertices, *A* and *B*, to form the equilateral triangle.

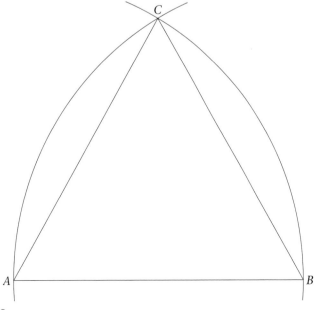

Perpendicular bisector

A **perpendicular bisector** cuts a line into two equal parts by a line that is at right angles to the original line. Any point on the perpendicular bisector is equidistant from the ends of the line.

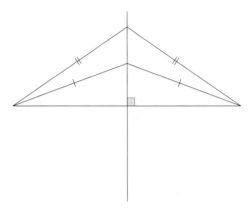

Keep your compass setting the same throughout.

Step 1
The line will already be drawn for you.

Step 2
Put your compass point on one end of the line and draw an arc above and below the given line. Make sure that your arc is quite long as shown.

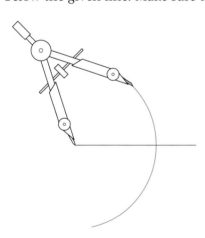

Step 3
Put the compass point at the other end of the line and draw another arc of equal radius to Step 2. Make sure that the arcs intersect as shown.

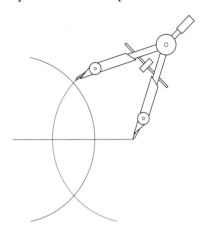

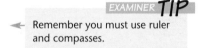

EXAMINER **TIP**

Remember you must use ruler and compasses.

Step 4
Use a ruler to complete the perpendicular bisector construction.

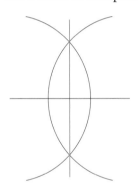

Angle bisector

An **angle bisector** cuts the angle into two equal angles. Any point on the angle bisector is the same distance from both lines forming the angle.

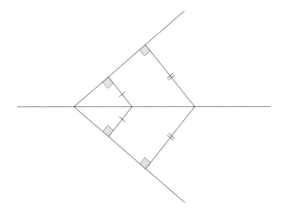

Keep your compass setting the same throughout.

Step 1
The angle will already be drawn for you.

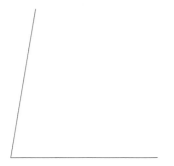

base line

Step 2
Put your compass point on the angle that you are bisecting and draw an arc.
Make sure that your arc cuts both lines as shown.

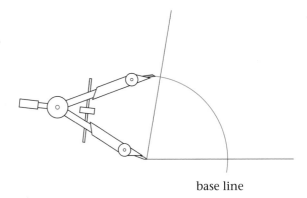

base line

Step 3
Now put the compass point at the point where your arc cuts the base line and
draw another arc inside the angle that you are bisecting.

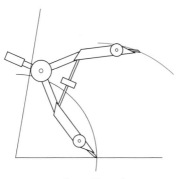

base line

Step 4

Now put the compass point at the point where your arc cuts the other line and draw another arc of equal radius to Step 3. Make sure that the arcs intersect as shown.

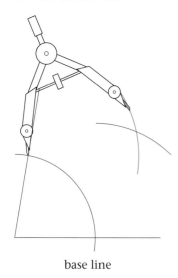

base line

Step 5

Use a ruler to complete the angle bisector construction.

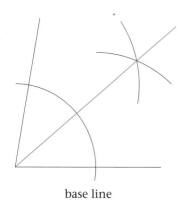

base line

EXAMINER **TIP**

Remember you must use ruler and compasses. No arcs means no marks.

Perpendicular from a point *P* to a line

Keep your compass setting the same between Steps 3 and 4. Your compass setting may change between Steps 2 and 3.

Step 1

The point and line will already be drawn for you.

P
×

Step 2

Make sure that your compasses are set so that they will make two clear intersections with the line from the point.

Put your compass point on the given point P and draw an arc cutting the line in two places.

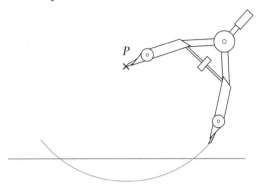

Step 3

Put the compass point at the first point of intersection with the line and draw another arc below the line.

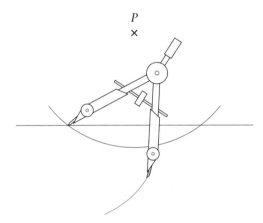

Step 4

Put the compass point at the other point of intersection with the line and draw another arc of equal radius to Step 3. Make sure that the arcs intersect as shown.

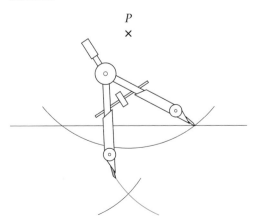

Step 5

Use a ruler to join the intersection of the arcs to the point *P* to complete the perpendicular from the point *P* to the given line construction.

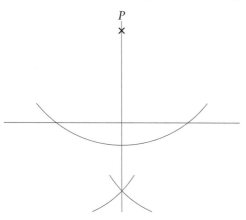

EXAMINER *TIP*

Remember you must use ruler and compasses.

90° at the end of a line

Keep your compass setting the same between Steps 5 and 6.

Step 1
The line will already be drawn for you.

Step 2
Mark the end of the line that you are using with a vertical dash as shown.

Step 3
Extend the line using a ruler and pencil.

Step 4

Put your compass point on the vertical dash and make two intersections with the line, one on either side of the vertical dash.

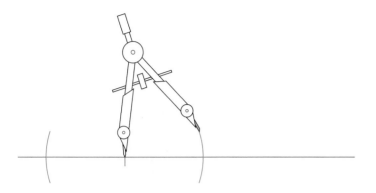

Step 5

Open your compasses wider. Put the compass point on the first point of intersection of the line and draw an arc above (or below) the line.

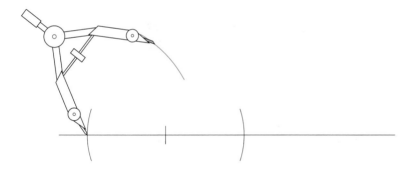

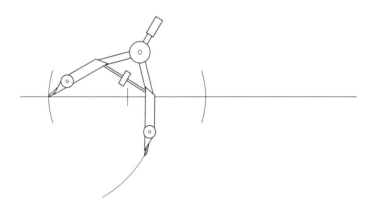

Step 6
Put the compass point at the other point of intersection of the line and draw another arc of equal radius to Step 5 above (or below) the line. Make sure that the arcs intersect as shown.

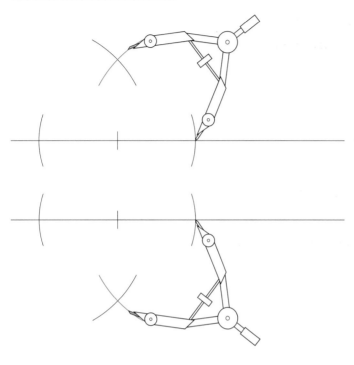

Step 7
Use a ruler to complete the perpendicular at the end of a line construction.

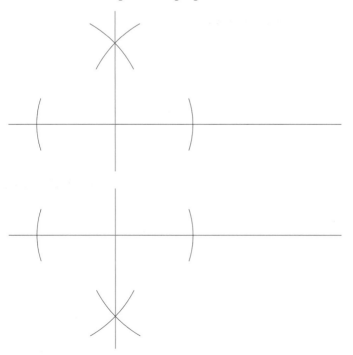

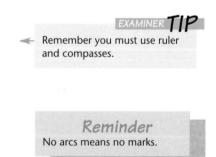

EXAMINER *TIP*
Remember you must use ruler and compasses.

Reminder
No arcs means no marks.

Practice questions

1   Using a ruler and compasses only, construct equilateral triangles with side lengths:

   a 4 cm   **b** 6 cm   **c** 7.5 cm   **d** 10 cm

   Check with a protractor that each angle in each triangle is 60°.

Practice exam questions

1   The scale diagram below shows a plan of a room.
    The dimensions of the room are 9 m and 7 m.
    Two plug sockets are fitted along the walls.
    One is at the point marked *A*. The other is at the point marked *B*.
    A third plug socket is to be fitted along a wall.
    It must be equidistant from *A* and *B*.
    Make a copy of the diagram. *Using ruler and compasses*, find the position of the new socket.
    Label it *C*.

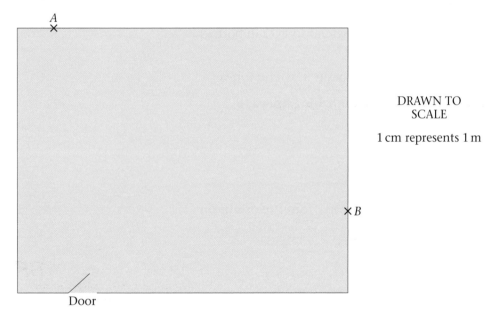

DRAWN TO
SCALE

1 cm represents 1 m

[AQA (NEAB) 2000]

2   The diagram shows a sketch of a right-angled triangle.

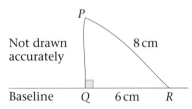

PR = 8 cm   *QR* = 6 cm

Using ruler and compasses only:
Construct an accurate drawing of the triangle *PQR*.

[AQA (NEAB) 2000]

29 Loci

A **locus** is a path made up of an infinite number of points that join together to form a line or curve. You may either be asked to draw a locus or to describe in words the shape of a locus.

The locus of all the points that are the same distance from a fixed point is a circle.

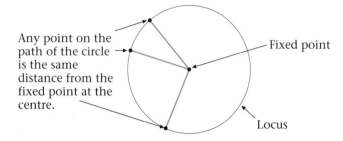

Any point on the path of the circle is the same distance from the fixed point at the centre.

Fixed point

Locus

Key vocabulary

Line: A straight line that is considered to extend forever in both directions.

Line segment: A line that has definite ends, i.e. it does not extend forever.

Equidistant: If two points are equidistant from something, they are exactly the same distance from it.

Example 29.1

Draw the locus of all of the points that are 1 cm away from the origin on a graph.

Solution

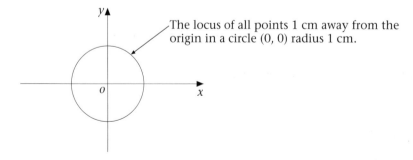

The locus of all points 1 cm away from the origin in a circle (0, 0) radius 1 cm.

Example 29.2

Draw the locus of all of the points that are exactly 2 cm away from a given line.

Solution

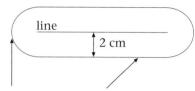

The locus has 2 semicircular ends and has straight sides. All are exactly 2 cm from the given line.

Locus of more than one line

The locus of all the points that are equidistant from two straight lines that form an angle is the angle bisector.

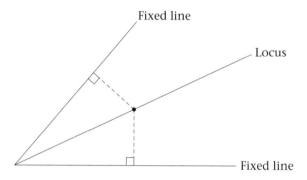

> **Reminder**
> Equidistant means 'the same distance from'.

Example 29.3

A scale drawing of a rectangular garden is shown.

Copy the rectangle and make an accurate scale drawing of a locus around the outside of the garden which is always 5 m from the garden.

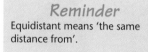

Solution

It is important to realise that the locus at the corners will be quarter circles.

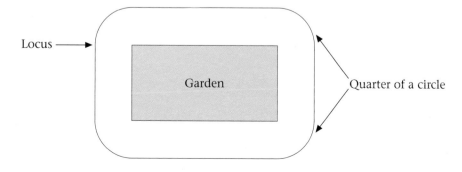

Example 29.4

The diagram shows a garden.
The owner wishes to plant a tree. It must be at least 1 metre from the edge of the garden and 1 metre from the rockery and the pond.

Show on the diagram the area where the tree can be planted.

1 cm represents 1 m

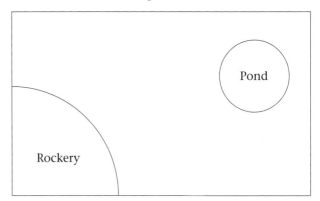

Solution

1 cm represents 1 m

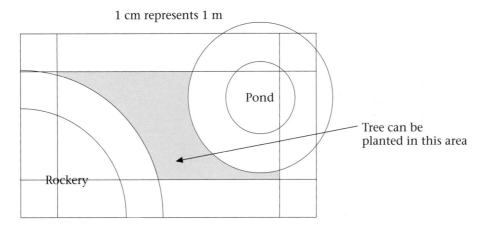

Tree can be planted in this area

Example 29.5

Draw the locus that is twice as far from the *y*-axis as it is from the *x*-axis.

Solution

If the point is always twice as far from the *y*-axis as it is from the *x*-axis then this is the situation.

This means that the gradient of the line (which is the locus of the point) is $\dfrac{h}{2h} = \dfrac{1}{2}$.

Since the line passes through the origin it has no *y*-intercept.

The equation of the line is $y = \dfrac{1}{2}x$.

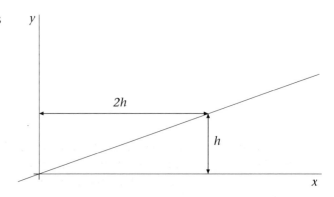

Example 29.6

Look at the diagram below. A goat is tethered in a field to a rectangular fence.

Copy this diagram and draw on it the locus points that show the boundary of where the goat can go.

You must label the distances of the locus from the fence.

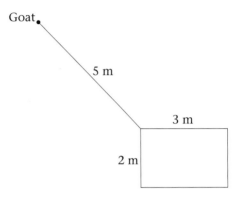

Solution

First imagine that the goat is walking around anticlockwise. It is keeping the rope taut. It will be able to walk a circle with radius 5 m until the rope is in line with the left-hand side of the rectangle.

Draw this circle on your diagram and mark the distance of the large circle from the fence.

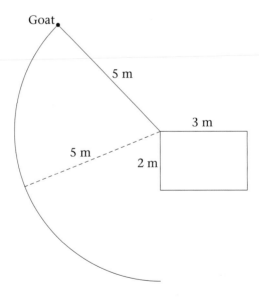

Now consider the point where the goat walks past the bottom left-hand corner. As the rope is stretched around the corner, a new circle will be drawn with the centre as the bottom left-hand corner and the radius of 3 m (length of the rope – length of the side = 5 – 2 = 3 m).

Draw this circle marking the distance between the fence and the locus.

As the bottom length of the rectangle is the same as the radius of the circle, the goat will not be able to go beyond the end of this side so the locus ends there.

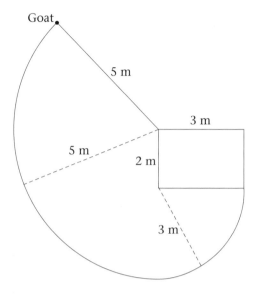

Now go back up to the top side of the rectangle and consider the goat walking clockwise. It will continue its circle of radius 5 m until it is in-line with the top side of the rectangle. After this it will start a new circle with the centre as the top right-hand corner and radius 2 m (length of rope − length of top side = 5 − 3 = 2 m).

As the length of the left-hand side of the rectangle is 2 m, the goat will not be able to go beyond the bottom right-hand corner.

Draw these last two circles and add the distances between the locus and the fence. You have now completed drawing the locus that shows where the goat can walk.

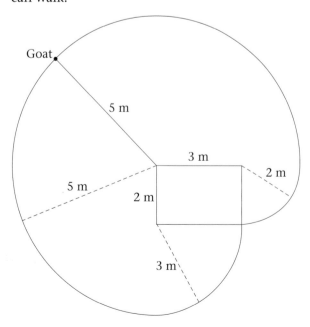

Practice questions

1   For each shape copy the diagram and draw a locus around the outside of
    the shape which is always 1 centimetre away from the shape.

a 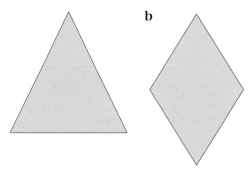   **b**   **c**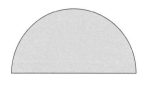

2   The diagram shows an L-shaped room.
    The only plug in the room is shown on the diagram.
    A radio has a lead of 2.5 metres.

    Copy the diagram and mark the positions on the
    drawing where the radio can be played.

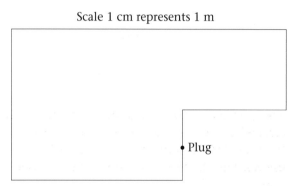

Scale 1 cm represents 1 m

Plug

Practice exam questions

1   The diagram shows a
    quadrilateral *ABCD*.

    a On a copy of the diagram
    draw the locus of points that
    are the same distance from
    AD and *DC*.
    b Shade the region inside the
    quadrilateral which is more
    than 5 cm from
    B and nearer to *DC* than to
    AD.

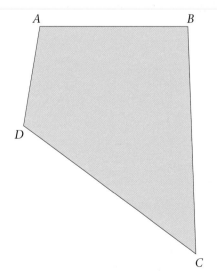

[AQA (SEG) 2001]

225

2 The plan shows the landing area, *ABCD*, for a javelin event.
AD is the throwing line.
The arc *BC* is drawn from the centre *X*.
The plan has been drawn to a scale of 1 cm to 5 m.

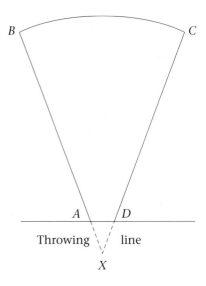

The landing area is fenced off in front of the throwing line.
The position of the fence is always 10 m from the boundaries *AB*, *BC* and
CD of the landing area.
Draw accurately the position of the fence on a copy of the plan.   [AQA (SEG) 2000]

3 The diagram shows a penguin pool at a zoo.
It consists of a right-angled triangle and a semicircle.
The scale is 1 cm to 1 m.

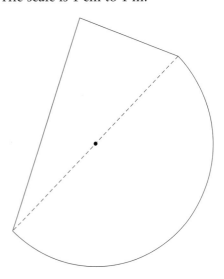

A safety fence is put up around the pool. The fence is always 2 m from the
pool.
Draw accurately the position of the fence on a copy of the diagram.   [AQA (SEG) 1998]

30 Bearings and scale drawings

Compass points

It is expected that you already know the eight compass directions. The main compass directions are called North, East, South and West.

North-west (NW) is exactly halfway between North and West.

South-east (SE) is exactly halfway between South and East.

South-west (SW) is exactly halfway between South and West.

North-east (NE) is exactly halfway between North and East.

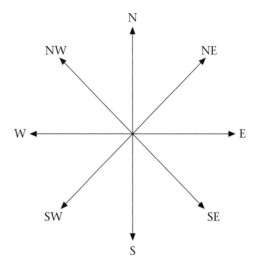

Three-figure bearings

Directions are often described using **three figure bearings**.

Three figure bearings are measured in degrees (using a protractor) from the North line in a clockwise direction.
Here are some examples:

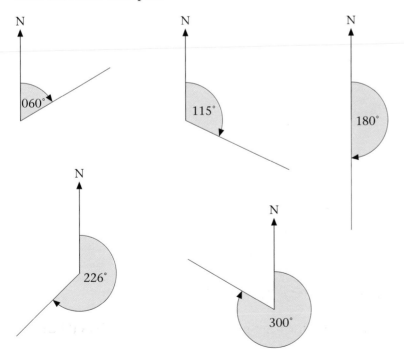

Practice questions 1

1   Use a protractor to draw accurate diagrams to represent these bearings.

   a 045°
   b 090°
   c 225°
   d 270°
   e 315°
   f 023°
   g 085°
   h 264°
   i   350°

2   Measure each of the following bearings giving your answer as a three figure bearing.

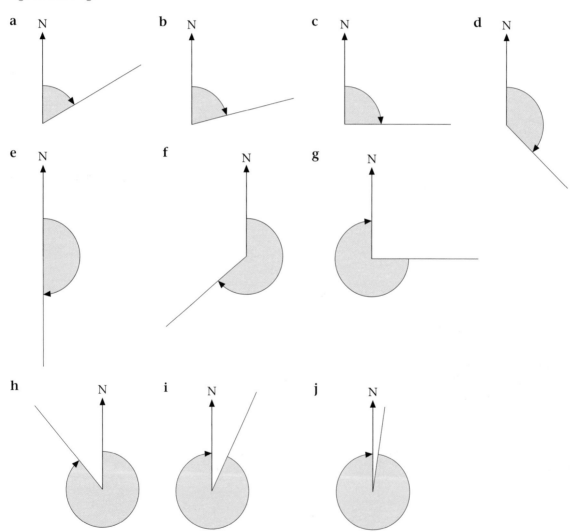

Drawing sketches and calculating bearings

You could be asked to calculate bearings from a diagram not drawn to scale. It is important that you do not measure angles if it is stated that the diagram is not drawn to scale or not drawn accurately.

Example 30.1

Harthill is on a bearing of 155° from Kiveton.

a Make a sketch of the positions of Harthill and Kiveton.

b Work out the bearing of Kiveton from Harthill.

Solution

a

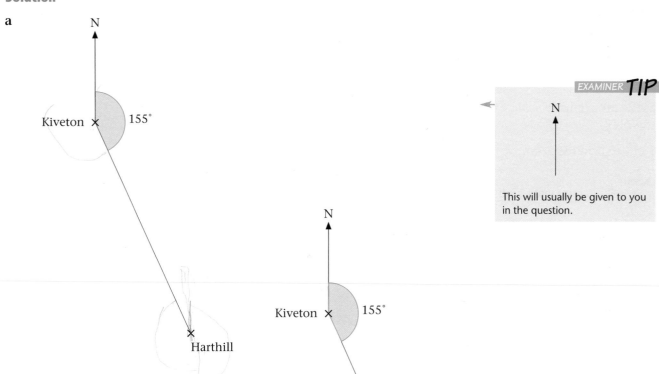

EXAMINER **TIP**

N

This will usually be given to you in the question.

b A person at Harthill faces North.
The person needs to turn 155° to face the opposite direction to Kiveton.
The person would then need to turn another 180° clockwise to face Kiveton.
So the bearing of Kiveton from Harthill is 155° + 180° = 335°.

Scale drawings

You will need to know how to construct accurate **scale drawings** (using a protractor) and you may also have to work out bearings by measuring angles on a diagram.

The scale of any diagram will usually be given in the form 1 cm represents 5 km, for example. As an alternative the scale may be given in ratio form, for example 1 : 500 000.

5 km = 5000 m
      = 500 000 cm
So 1 cm : 5 km is the same as 1 : 500 000.

Example 30.2

Rotherford (R) is on a bearing of 214° from Castleham (C).

The distance from Rotherford to Castleham is 7 km.

Make an accurate scale drawing to show the position of Rotherford on the map.

1 cm represents 2 km.

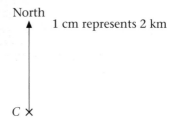

Solution

First calculate the scaled distance between Rotherford and Castleham.

The scale is 1 cm representing 2 km.

This means that 7 km is represented by 7 ÷ 2 = 3.5 cm on the map.

Now draw a line on the bearing 214°. Measure an angle of 214° from C in a clockwise direction from the North line on the map as shown below. Draw a line 3.5 cm long, as this is the scaled distance between the two places.

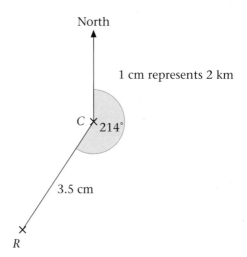

Bearings and trigonometry

Some questions that test bearings may also have a part that tests your knowledge of trigonometry.

Example 30.3

The distance from *P* to *Q* is 10 km.

Q is on a bearing of 140° from *P*.

How far is *P* North of *Q*?

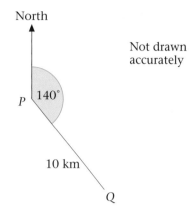

Not drawn accurately

Solution

Method A (using the sine ratio)

The distance required (marked *x* on the diagram below) can be found using the right-angled triangle.

$$\sin 50° = \frac{\text{opposite}}{\text{hypotenuse}}$$

$$\sin 50° = \frac{x}{10}$$

$$10 \times \sin 50° = x$$

$$7.66 = x$$

So *P* is approximately 7.7 km North of *Q*.

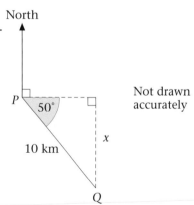

Not drawn accurately

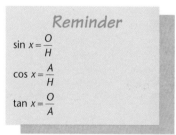

Method B (using the cosine ratio)

The distance required (marked *x* on the diagram below) can be found using the right-angled triangle.

$$\cos 40° = \frac{\text{adjacent}}{\text{hypotenuse}}$$

$$\cos 40° = \frac{x}{10}$$

$$10 \times \cos 40° = x$$

$$7.66 = x$$

So *P* is approximately 7.7 km North of *Q*.

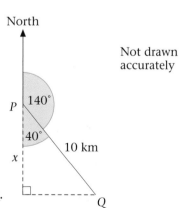

Not drawn accurately

Practice exam questions

1   The map of an island is shown.
    The map has been drawn to a scale of 5 cm to 4 miles.

    a The port is at *A* and the airport at *B*.
        Use the map to find the distance *AB* in miles.
    b The Hotel Central is equidistant from *A* and *B* on a
        bearing of 150° from *A*.
        On a copy of the diagram draw loci to represent this
        information.
        Mark with a cross, the position of the hotel on the
        map.

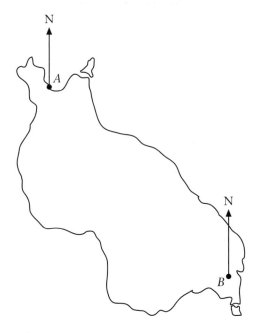

[AQA (SEG) 1999]

2

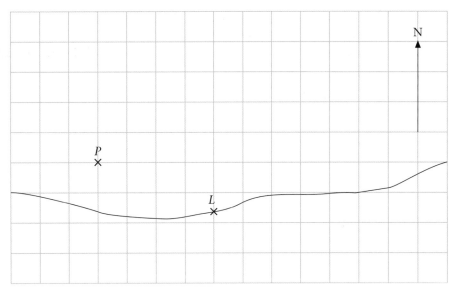

The map shows the position of a ship, *P*, and a lighthouse *L*.

    a What is the bearing of *P* from *L*?
    b Another ship *Q* is due North of *L*.
        Q is on a bearing of 055° from *P*.
        On a copy of the diagram mark clearly the position of *Q*.

[AQA (NEAB) 2001]

3 The diagram shows an accurate plan of a race.

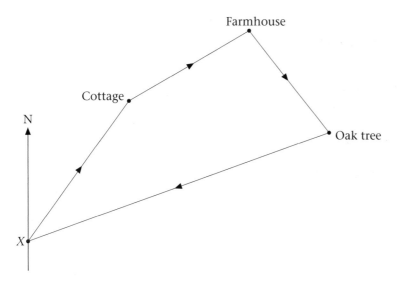

a The start and finish of the race is at *X*.

    i What is the bearing of the cottage from *X*?

    ii What is the bearing of *X* from the oak tree?

b The plan has been drawn using a scale of 1 mm to represent 10 m.
Use the map to estimate the length of the race in kilometres.
Give your answer to the nearest tenth of a kilometre.   [AQA (SEG) 1999]

4 The map shows two roads which meet at *O*.
The map has been drawn to a scale of 1 cm to 200 m.

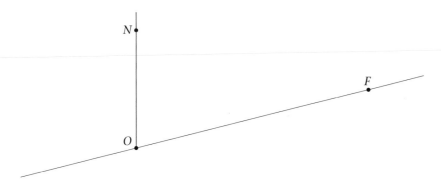

There is a farmhouse at *F*.

a Use the map to calculate the distance *OF* in kilometres.

b *N* is due North of *O*.
A haystack is on a bearing of 120° from *O* and 180° from *F*.
Mark, with a cross, the position of the haystack on a copy of the map.

c Sanjay cycles from the farmhouse.
He travels 5 km in 12 minutes.
Calculate his average speed in kilometres per hour.   [AQA (SEG) 2000]

5 The map shows the positions of Huddersfield, Emley and Rotherham.
The scale is 1 cm represents 2.5 km.

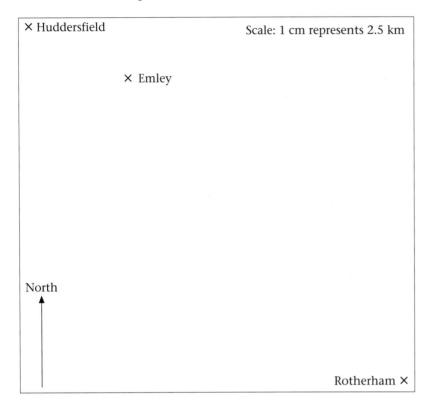

× Huddersfield   Scale: 1 cm represents 2.5 km

× Emley

North

Rotherham ×

a Use the map to calculate the actual straight line distance from
Huddersfield to Rotherham.
Show your working.
b Penistone is 20 km from Rotherham.
Penistone is also due South of Emley.
Mark the position of Penistone on a copy of the map.   [AQA (NEAB) 2001]

6 The bearing of Cardiff from Leeds is 200°.
Cardiff is 209 miles from Leeds.

a Show this information in a copy of the sketch.

North

Leeds •

Cardiff •

b The diagram below shows the positions of Leeds and Newcastle.

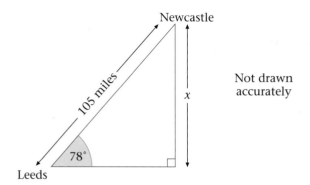

Not drawn accurately

Calculate how far South Leeds is from Newcastle, the distance marked *x* on the diagram.
Give your answer to an appropriate degree of accuracy.

[AQA (NEAB) 2000]

7 A piece of land is bounded by three straight roads *PQ*, *QR* and *RP*, as shown.

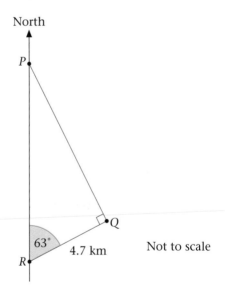

Not to scale

P is due North of *R*.
Q is on a bearing of 063° from *R*.
PQR is a right angle.

a What is the bearing of *P* from *Q*?
   The length *QR* is 4.7 km.
b Calculate the area of the piece of land.

[AQA (SEG) 1999]

31 Recognising graphs

You have to be able to recognise various graphs, giving the equation when they are already sketched or sketching the graphs from their equations. The most basic form of each graph is shown below.

Linear graphs

In Chapter 15 you looked at linear (straight line) graphs. These have equations of the form $y = mx + c$.

You need to recognise the graph of $y = x$.

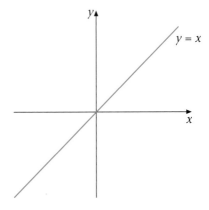

You also need to recognise the graph of $y = -x$.

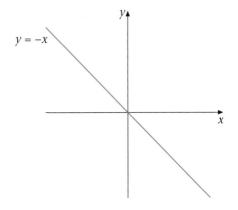

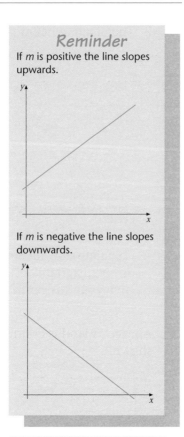

Reminder

If m is positive the line slopes upwards.

If m is negative the line slopes downwards.

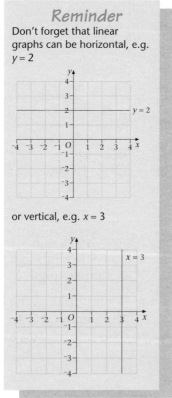

Reminder

Don't forget that linear graphs can be horizontal, e.g. $y = 2$

or vertical, e.g. $x = 3$

Quadratic graphs

A quadratic graph has an equation of the form $y = ax^2 + bx + c$ where $a \neq 0$.

All quadratic graphs are curves with the same basic shape:

If $a > 0$:

If $a < 0$:

The simplest quadratic graph is $y = x^2$.

You are also expected to be able to draw or recognise the graph of $y = -x^2$.

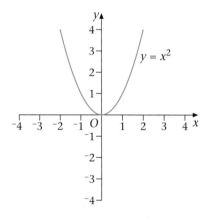

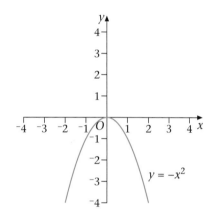

Cubic graphs

You are expected to be able to draw or recognise the graph of $y = x^3$.

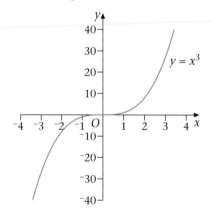

You may also be asked to plot, using a table of values, other cubic graphs such as $y = -x^3$.

x	-3	-2	-1	0	1	2	3
y	27	8	1	0	-1	-8	-27

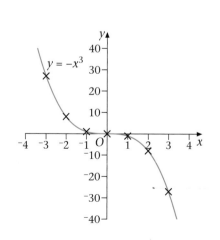

Reciprocal graphs

You are expected to be able to draw or recognise the graph of $y = \dfrac{1}{x}$.

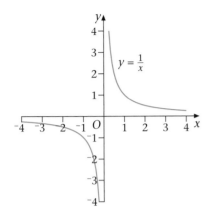

You may also be asked to plot, using a table of values, other reciprocal graphs such as $y = \dfrac{12}{x}$ for positive values of x.

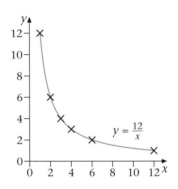

x	1	2	3	4	6	12
y	12	6	4	3	2	1

Example 31.1

On the same grid, sketch and label the following graphs.

a $y = x$ b $y = x + 3$ c $y = 2x$

Solution

a Firstly, draw the line $y = x$.

b $y = x + 3$ is parallel to $y = x$ but intercepts the y-axis at $(0, 3)$.

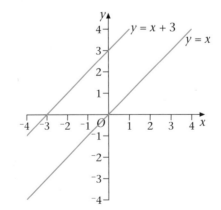

c $y = 2x$ is steeper than $y = x$ but also passes through the origin.

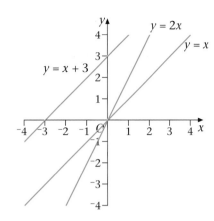

Example 31.2

On the same grid, sketch and label the following graphs.

a $y = x^2$ **b** $y = x^2 - 3$ **c** $y = -x^2$

Solution

a Firstly, draw the curve $y = x^2$.

b $y = x^2 - 3$ is a translation of the curve $y = x^2$ by 3 units in the negative y direction.
The curve intercepts the y-axis at $(0, -3)$.

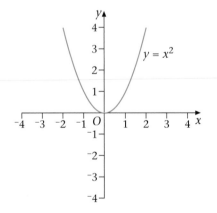

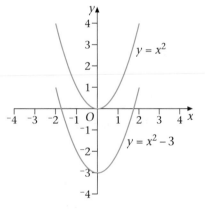

c $y = -x^2$ is a reflection of $y = x^2$ in the x-axis.
It also passes through the origin as there is no constant term.

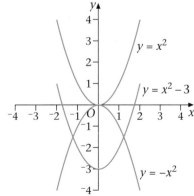

Example 31.3

On the same grid, sketch and label the following graphs.

a $y = x^3$

b $y = x^3 + 2$

c $y = x^3 - 4$

Solution

a Firstly, draw the curve $y = x^3$.

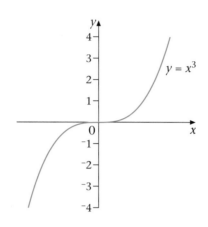

b $y = x^3 + 2$ is a translation of the curve $y = x^3$ by 2 units in the positive y direction.
The curve intercepts the y-axis at $(0, 2)$

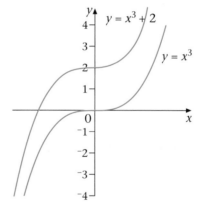

c $y = x^3 - 4$ is a translation of the curve $y = x^3$ by 4 units in the negative y direction.
The curve intercepts the y-axis at $(0, -4)$.

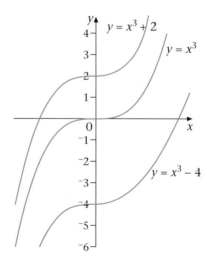

Practice question

1 Match each of the following equations with its graph.

    a $y = -2$
    b $y = x - 2$
    c $y = -x^2$
    d $x = 2$
    e $y = 2x$

A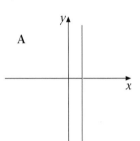
B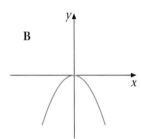
C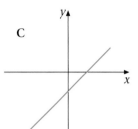

D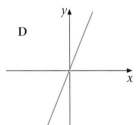
E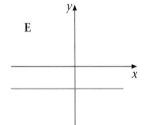

Practice exam questions

1 Each of the graphs below represents one of the following equations.

    A $y = x^2$
    B $x = 1$
    C $y = 1$
    D $y = x$
    E $y = \dfrac{1}{x}$

Write down the equation represented by each graph.

a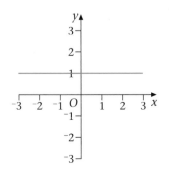
b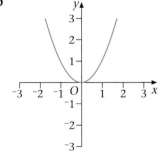

c 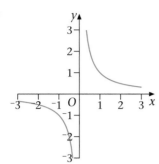   d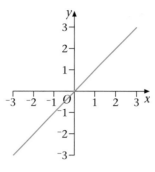

[AQA (NEAB) 2001]

2   Match three of these equations with the graphs shown below.

$y = 5 - x$

$y = 5 - x^2$

$y = 5x^2$

$y = x + 5$

$y = 5x$

a 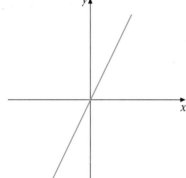   b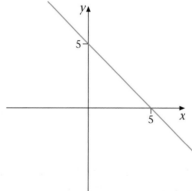

c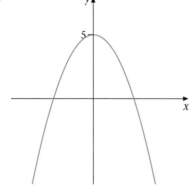

[AQA (SEG) 2000]

3 The diagram shows a sketch of the graph $y = 2x + 3$.

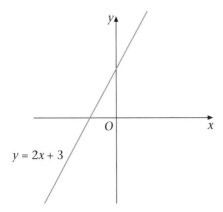

On a copy of the diagram, draw and label sketch graphs of:

a $y = 3$
b $y = x + 3$.

[AQA (NEAB) 2002]

32 Dimensional analysis

Dimensional analysis is the study of expressions or formulae to decide whether they represent **length**, **area**, **volume** or **none** of these.

In every question you will be told that each letter in the expression represents a length.

Here you look at a rectangle of length a and width b. It does not matter in these questions whether the lengths are in mm, cm, m or km.

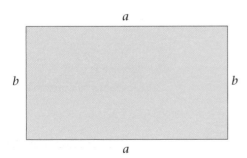

The perimeter of the rectangle is $a + b + a + b$ which simplifies to $2a + 2b$.

a is a length, b is a length and if you add lengths together, as in this case, the answer (perimeter) is also a length.

This shows that: **length + length = length**

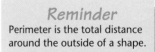

Reminder
Perimeter is the total distance around the outside of a shape.

243

Length has one dimension.

This means that all the following expressions represent lengths.

a

$2a$

$a + b$

$a + b + c$

$2a + 2b$

$2\pi r$

> **Reminder**
> Circumference of a
> circle = $2\pi r$

Look again at the rectangle.

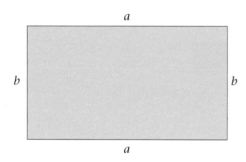

The area of the rectangle is calculated using this formula:

Area = length × width

So in this case   Area = $a \times b$ or ab

This shows that: **length × length = area**

Area has two dimensions.

This means that all the following expressions represent areas.

> **Reminder**
> ab is the same as $a \times b$

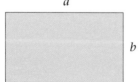

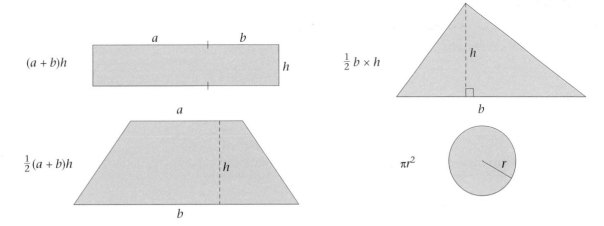

$(a + b)h$

$\frac{1}{2}(a + b)h$

$\frac{1}{2} b \times h$

πr^2

Now look at a cuboid with dimensions a, b and c.

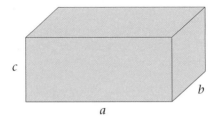

Reminder
A cuboid is a rectangular box.

The volume of the cuboid is calculated using this formula:

Volume = length × width × height

So in this case   Volume = $a \times b \times c$ or abc

This shows that: **length × length × length = volume**

Volume has three dimensions.

This means that all the following expressions represent volumes.

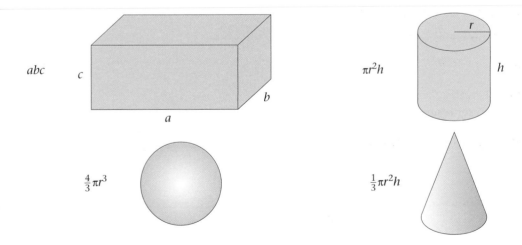

abc

$\pi r^2 h$

$\frac{4}{3}\pi r^3$

$\frac{1}{3}\pi r^2 h$

Now you will look at how to decide the answer without having to draw a diagram.

Here are the facts to remember: Length = L, Area = A, Volume = V

$L \times L = A$
$L \times L \times L = V$

Any numbers in the expression, including π, can be ignored.

Expression	Numbers	Dimensions	Answer	Number of dimensions
a		L	Length	1
$2a$	2	L	Length	1
$a + b$		$L + L$	Length	1
ab		$L \times L$	Area	2
$\frac{1}{2}(a+b)h$	$\frac{1}{2}$	$(L + L) \times L = L \times L$	Area	2
$\pi r^2 h$	π	$L \times L \times L$	Volume	3
$a^2 + b^2$		$(L \times L) + (L \times L) = A + A$	Area	2
$a^2 + b$		$(L \times L) + L = A + L$	None	0
$\dfrac{xy^2 z}{w}$		$\dfrac{L \times L \times L \times \cancel{L}}{\cancel{L}}$ $= L \times L \times L$ $= V$	Volume	3

Note: You cannot add expressions with different dimensions, e.g. $a^2 + b$ represents area and length which cannot be added. In these cases, there can be no dimension analysis, so the answer is none.

Practice question

1   w, x, y and z each represent lengths.
State whether each expression represents length, area, volume or none of these.

a $w + y$
b $x^2 z$
c $2(w + x + y + z)$
d y^2
e $x^2 + y^2$
f $\dfrac{wy}{z}$
g $w^3 + 2xy$
h $4\pi r^2$
i $\frac{1}{2}(x + y + z)$
j $3x + z^2$

Practice exam questions

1 The letters a, b, h and r represent lengths.
Which of the following formulae could represent volume?

$$\pi r(a^2 + b^2) \qquad \pi rh \qquad \pi r\sqrt{a^2 + h^2} \qquad \frac{\pi r^2}{3}(h + 2r)$$

[AQA (SEG) 2000]

2 In the following expressions r, a and b represent lengths.
For each expression state whether it represents
a *length*,
an *area*,
a *volume*,
or *none* of these.

a πab

b $\pi r^2 a + 2\pi r$

c $\dfrac{\pi ra^3}{b}$

[AQA (NEAB) 2000]

3 The following formulae represent certain quantities connected with containers, where a, b and c are dimensions.

$$\pi a^2 b \qquad 2\pi a(a + b) \qquad 2a + 2b + 2c \qquad \tfrac{1}{2}(a + b)c \qquad \sqrt{a^2 + b^2}$$

a Which of these formulae represent area?
b Which of these formulae represent volume?

[AQA (SEG) 1999]

4 The diagram shows a prism.

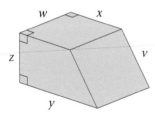

z w x v y Not to scale

The following formulae represent certain quantities connected with the prism.

$$wx + wy \qquad \frac{1}{2}z(x + y)w \qquad \frac{z(x + y)}{2} \qquad 2(v + 2w + x + y + z)$$

a Which of these formulae represents length?
b Which of these formulae represents volume?

[AQA (SEG) 2000]

33 Similarity

Two shapes are **similar** if one shape is an enlargement of the other. The size of the enlargement is called the **scale factor**.

When two shapes are similar the corresponding *angles are equal* and the lengths of the corresponding *sides are in the same ratio*.

These are similar shapes.

The angles marked $x°$ are corresponding angles because they are located in the same position in each shape.

These shapes are also similar:

Two rectangles are similar if the corresponding *sides are in the same ratio*.

The corresponding angles are always equal as they are all 90°.

Not drawn
accurately

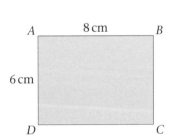

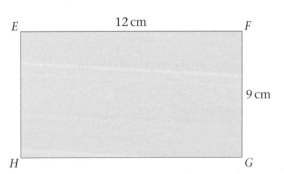

Rectangle *ABCD* is similar to rectangle *EFGH* because the corresponding angles are equal and the ratio of length to width of each rectangle is equal.

This means $\dfrac{8}{6} = \dfrac{12}{9}$.

Rectangle *EFGH* is an enlargement of rectangle *ABCD* by a scale factor of $\dfrac{9}{6} = 1.5$.

Two triangles are similar when their corresponding sides are in the same ratio and their corresponding angles are equal. Consider triangles *ABC* and *DEF*.

Not drawn accurately

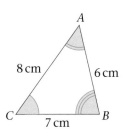
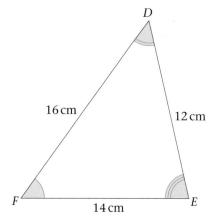

Triangles *ABC* and *DEF* are similar because their corresponding sides are in the same ratio.

Side *AC* corresponds to side *DF*.

Side *AB* corresponds to side *DE*.

Side *BC* corresponds to side *EF*.

The length of each side of triangle *ABC* is half the length of the corresponding side of triangle *DEF*, i.e. $\dfrac{AC}{DF} = \dfrac{AB}{DE} = \dfrac{BC}{EF} = \dfrac{1}{2}$.

Alternatively, you can say the length of each side of triangle *DEF* is twice the length of the corresponding side of triangle *ABC*, i.e. $\dfrac{DF}{AC} = \dfrac{DE}{AB} = \dfrac{EF}{BC} = \dfrac{2}{1}$.

This similarity relationship can be written either way.

Also:

$\angle A = \angle D$
$\angle B = \angle E$
$\angle C = \angle F$

These are the corresponding angles.

Triangle *DEF* is an enlargement of triangle *ABC* with scale factor $\dfrac{14}{7} = 2$.

Using the scale factor method

When two shapes are similar, one is a scale factor enlargement of the other. You can find the scale factor of the enlargement and use it to calculate missing lengths of sides.

Example 33.1

The two shapes are similar.

Find the value of:

a x

b y

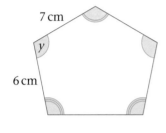

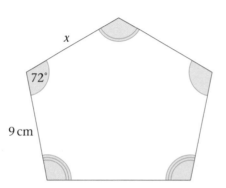

Not drawn accurately

Solution

a The two shapes are similar. The enlargement scale factor of the lengths is
$\dfrac{9}{6} = 1.5.$
So the lengths of the sides of the larger shape are 1.5 times as big as the lengths of the sides of the smaller shape.
Then $x = 1.5 \times 7$
$= 10.5$ cm

b The corresponding angles are equal in similar shapes and so $y = 72°$.

Example 33.2

The two shapes are similar.

Find x.

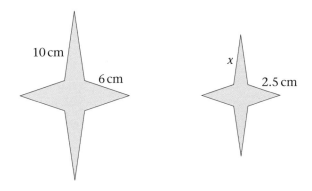

10 cm

6 cm

x

2.5 cm

Not drawn
accurately

Solution

The two shapes are similar.

The larger shape has sides which are a scale factor enlargement of the sides of
the smaller shape.

The scale factor is $\dfrac{6}{2.5} = 2.4$.

So the lengths of the sides of the larger shape are 2.4 times bigger than the
lengths of the sides of the smaller shape.

Or the lengths of the sides of the smaller shape are 2.4 times smaller than the
lengths of the sides of the larger shape.

You are asked to find x, the length of a side of the smaller shape.

$$x = \frac{10}{2.4}$$
$$= 4\tfrac{1}{6} \text{ or } 4.1666 \dots \text{ cm}$$
$$= 4.2 \text{ cm (1 d.p.)}$$

Using similarity

The similarity relationship enables you to find the sides of shapes or the angles of shapes which are similar to each other.

Example 33.3

These rectangles are similar.

Calculate the length, x, of rectangle *EFGH*.

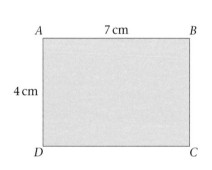

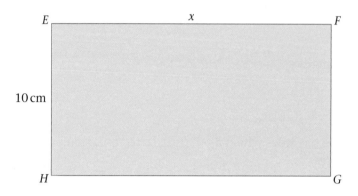

Not drawn
accurately

Solution

Method A (using similarity)
You are given that the rectangles are similar so the ratio of the corresponding sides are equal.

This means that:
$$\frac{EH}{AD} = \frac{EF}{AB}$$

Substituting the lengths of the sides into this equation:
$$\frac{10}{4} = \frac{x}{7}$$

Multiplying both sides by 7 gives:
$$7 \times \frac{10}{4} = x$$
$$17.5 \text{ cm} = x$$

Method B (scale factor method)
When two shapes are similar, one is a scale factor enlargement of the other. You can find the scale factor of the enlargement and use it to calculate missing lengths of sides.

Rectangle *EFGH* is an enlargement of rectangle *ABCD*.

The scale factor of the enlargement is $\frac{10}{4} = 2.5$.

So x must be $2.5 \times 7 = 17.5$ cm.

Example 33.4

Triangles *ABC* and *DEF* are similar.

Calculate the length *y*.

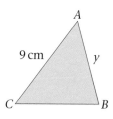

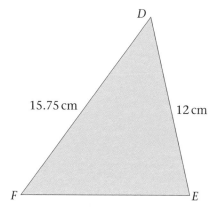

Not drawn
accurately

Solution

Method A (using similarity)
Triangles *ABC* and *DEF* are similar so their corresponding sides are in the same ratio.

Linking up the corresponding pairs of sides:

$$\frac{AC}{DF} = \frac{AB}{DE}$$

Substituting the lengths of the sides in this equation:

$$\frac{9}{15.75} = \frac{y}{12}$$

Multiplying both sides by 12 gives:

$$12 \times \frac{9}{15.75} = y$$

$$6.86 \text{ cm} = y$$

EXAMINER **TIP**
Always try to place the side you want to find on the top of the fraction. Here you wanted to find y, so AB was placed on the top of the fraction $\frac{AB}{DE}$.

Method B (scale factor method)
Triangle *DEF* is an enlargement of triangle *ABC*.

The scale factor of the enlargement is $\frac{15.75}{9} = 1.75$.

The lengths of the sides of triangle *DEF* are 1.75 times *larger* than the corresponding sides of triangle *ABC*.

The lengths of the sides of triangle *ABC* are 1.75 times *smaller* than the corresponding sides of triangle *DEF*.

In order to find *y* you have to divide side *DE* by 1.75:

$$y = \frac{12}{1.75}$$

$$= 6.86 \text{ cm}$$

EXAMINER **TIP**
Be careful. Make sure you know when you should multiply and when you should divide by the scale factor.

Example 33.5

The following shapes are similar.

Find the values of x, y and z.

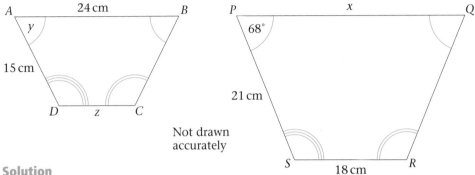

Not drawn accurately

Solution

Trapezium $ABCD$ is similar to trapezium $PQRS$, so the corresponding sides are in the same ratio and the corresponding angles are equal.

First find x, which is the length of side PQ. Write out the similarity relationship with PQ on the top of the fraction and connect it to another corresponding pair of sides whose values you know:

$$\frac{PQ}{AB} = \frac{PS}{AD}$$

Substitute the lengths of the sides in this equation:

$$\frac{x}{24} = \frac{21}{15}$$

Multiply both sides by 24:

$$x = \frac{21}{15} \times 24$$

$$= 33.6 \text{ cm}$$

Now find y.

Since the two shapes are similar the corresponding angles are equal.

In these shapes

$\angle A$ corresponds to $\angle P$

so $y = 68°$.

Finally find z, which is the length of side CD. Write out the similarity relationship with CD on the top of the fraction and connect it to another corresponding pair of sides whose values you know:

$$\frac{CD}{RS} = \frac{AD}{PS}$$

Substitute the lengths of the sides in this equation:

$$\frac{z}{18} = \frac{15}{21}$$

Multiply both sides by 18:

$$z = \frac{15}{21} \times 18$$

$$= 12.9 \text{ cm (1 d.p.)}$$

EXAMINER **TIP**

It is always better to use the values you are given rather than the values you have calculated yourself.

Proving similarity

To show that two triangles are similar you have to show that all three pairs of corresponding angles are equal or that the ratio of the corresponding sides is equal for all three pairs of sides.

Example 33.6

a Show that triangles *RST* and *LMN* are similar.

b Calculate the length *x*.

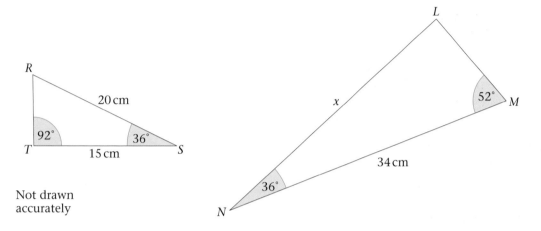

Not drawn accurately

Solution

a In this example you only know two of the sides of triangle *RST* so you can only show that the two triangles are similar by showing that the corresponding angles are equal.
In triangle *RST*:

$\angle R = 180° - 92° - 36°$
$ = 52°$

In triangle *LMN*:

$\angle L = 180° - 36° - 52°$
$ = 92°$

The corresponding angles are:

$\angle R = \angle M = 52°$
$\angle S = \angle N = 36°$
$\angle T = \angle L = 92°$

The angles in *LMN* are equal to the angles in *TRS* so these two triangles are similar.

b To find *x*, it is better to redraw the triangles so that the corresponding angles are in the same location.

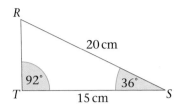

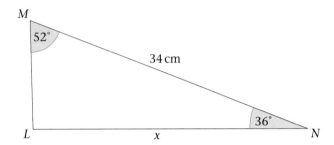

Write down the similarity relationship with *x* (side *LN*) on the top of the fraction:

$$\frac{LN}{TS} = \frac{MN}{RS}$$

Substitute the lengths of the sides into this equation:

$$\frac{x}{15} = \frac{34}{20}$$

Multiplying both sides of this equation by 15 gives:

$$x = \frac{34}{20} \times 15$$

$$= 25.5 \text{ cm}$$

Practice questions

1   In each part the two shapes are similar.
    Calculate the unknown lettered values.

a

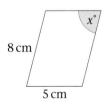

Not drawn
accurately

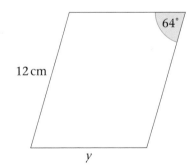

b

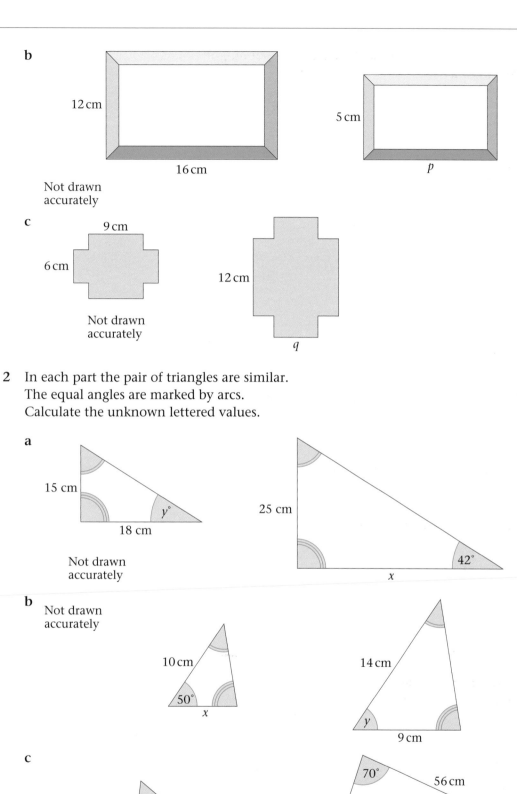

12 cm

16 cm

5 cm

p

Not drawn
accurately

c

9 cm

6 cm

12 cm

q

Not drawn
accurately

2   In each part the pair of triangles are similar.
The equal angles are marked by arcs.
Calculate the unknown lettered values.

a

15 cm

18 cm

y°

25 cm

42°

x

Not drawn
accurately

b   Not drawn
accurately

10 cm

50°

x

14 cm

y

9 cm

c

x

32 cm

70°

50°

z

70°

56 cm

60°

69 cm

y

Not drawn
accurately

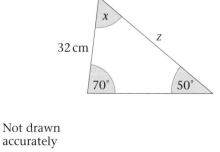

3 Each of these pairs of triangles are similar.
The equal angles are marked by arcs.
Calculate the lengths of the lettered sides.

a

10 cm

x

14 cm

y

16 cm

28 cm

Not drawn
accurately

b

p

15 cm

18 cm

q

15 cm

30 cm

Not drawn
accurately

c

Not drawn
accurately

a

25 cm

7 cm

b

42 cm

43.75 cm

4 A rectangular garden pond 3 m by 7 m has a 1 m path all the way around
the outside. Are the two rectangles similar? Explain your answer.

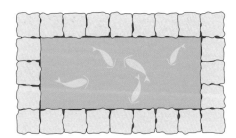

5 Triangles *PRS* and *PQT* are similar. Calculate the following lengths.

  a *TQ*
  b *PQ*

6 a Explain why triangles *LMP* and *NMO* are similar.
Angle *L* = angle *N*

  b *LM* = 10 cm
  MN = 16 cm
  MO = 14 cm
  Calculate the length of *PM*.

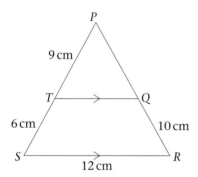

Not drawn accurately

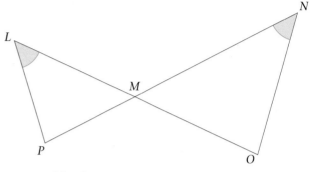

Not drawn accurately

Practice exam questions

1 These two shapes are similar.

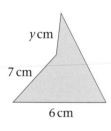

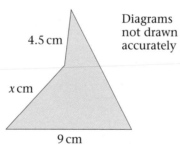

Diagrams not drawn accurately

  a Work out the value of *x*.
  b Work out the value of *y*.

[AQA (NEAB) 2002]

2 One of the earliest cameras was the pinhole camera.
Rays of light pass from the object through a small pinhole and form an
upside down image on the photographic film at the back of the camera, as
shown in the diagram.

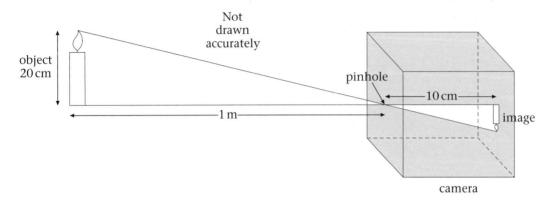

The photographic film is 10 cm from the pinhole.
An object 20 cm high is held 1 metre from the front of the camera.
How high will the image be? [AQA (NEAB) 2001]

3 In the diagram, *QR* is
parallel to *ST*.
a Explain why triangles
PQR and *PST* are
similar.
PR = 6 cm, *RT* = 4 cm
and *ST* = 12.5 cm.

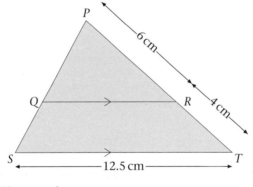

Not to scale

b Calculate the length of *QR*. [AQA (SEG) 2002]

4 Three triangles are joined together as
shown in the diagram.
DOA is a straight line.
OA = 2 cm and *OB* = 4 cm.
The three triangles are similar to each
other.
Calculate the length *AD*.

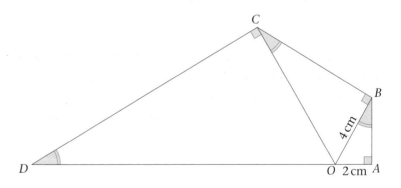

Not drawn
accurately

[AQA (NEAB) 2000]

34 Congruency

Two shapes are said to be **congruent** when one is exactly the same shape and same size as the other. Congruent shapes are identical to each other and one shape will fit exactly over the other shape.

These shapes are congruent.
They will all fit exactly over each other.

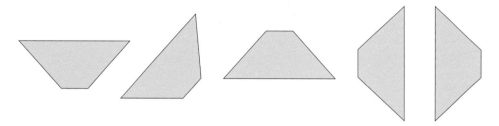

Example 34.1

Which two of the following shapes are congruent?

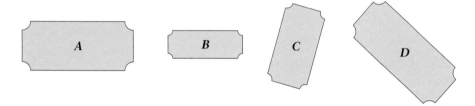

Solution

All of the shapes have the same basic style but only shapes A and D are the same shape and the same size. You say that shapes A and D are congruent. Shape D will fit exactly over shape A and vice-versa. Shapes B and C are different sizes to shapes A and D.

> **Reminder**
> Shapes which are enlargements of each other are said to be similar. See Chapter 33.

Congruent triangles

Two triangles are congruent when they are exactly the same shape and size. One triangle will fit exactly over the other triangle.

This means that congruent triangles must have three pairs of corresponding equal sides and three pairs of corresponding equal angles.

> *EXAMINER* **TIP**
> ← Corresponding sides are 'in the same position', i.e. opposite the same angle.

The markings on the triangles below indicate which sides and which angles are equal. For example, angle A = angle E and side AC = side ED.

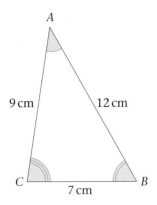
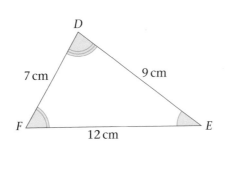

The triangles have three equal corresponding sides and three equal corresponding angles.

Triangle ABC is congruent to triangle EFD.
This is written $\triangle ABC \equiv \triangle EFD$.

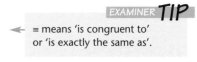

EXAMINER **TIP**

≡ means 'is congruent to' or 'is exactly the same as'.

You should order the letters of the second triangle to match the order in which you have written the corresponding angles in the first triangle. In the triangles above, the corresponding angles are:

$$\angle A = \angle E \qquad \angle B = \angle F \qquad \angle C = \angle D$$

So, using this order, you say triangle ABC is congruent to triangle EFD. Alternatively, you could say triangle DEF is congruent to triangle CAB, because the angles are still in corresponding order.

Example 34.2

Triangles *XYZ* and *RST* are congruent. Show how you can make a kite with these triangles.

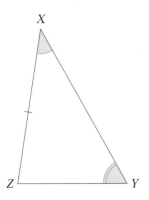
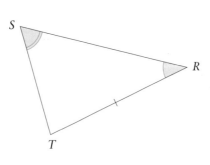

Solution

$XZ = RT$ (equal lengths indicated by marking on diagram)

$\angle X = \angle R$

$\angle Y = \angle S$

Place sides XZ and RT side by side as follows:

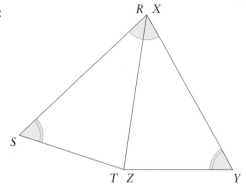

Practice questions

1   Write down which pairs of shapes are congruent.

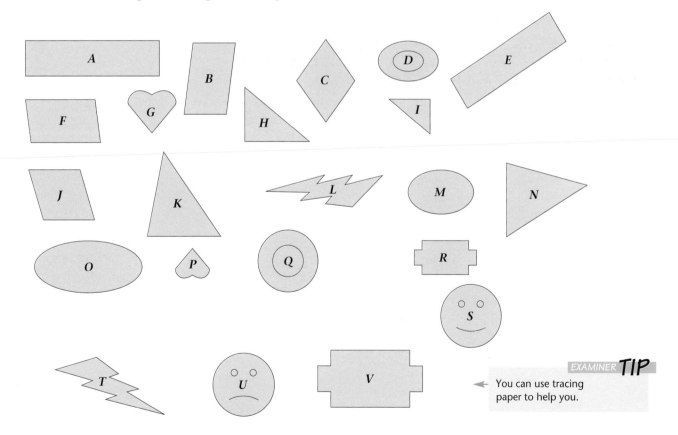

2   Triangles *PQR* and *LMN* each have angles of 68°, 49° and 63°.
Explain why these triangles may not be congruent.

3 Which pairs of triangles are congruent?

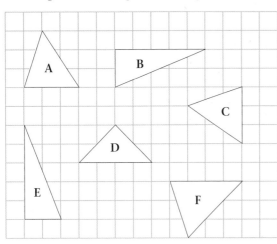

4 A square can always be cut into two congruent triangles as shown.
Show how three other shapes can be cut into two congruent triangles.

Practice exam questions

1 A teacher sketches five triangles. Which two of these triangles are congruent? Give a reason for your answer.

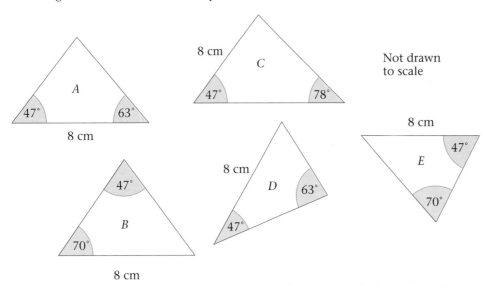

Not drawn to scale

[AQA (NEAB) 1999]

2 The two triangles shown below are congruent.

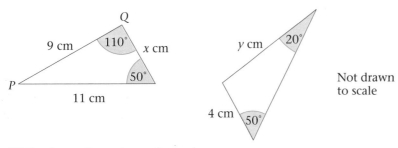

Not drawn to scale

Write down the values of *x* and *y*.

[AQA (SEG) 2002]

35 Simultaneous equations

$x + y = 7$ is an equation with two unknown quantities x and y. There are an infinite number of possible values for x and y which will solve this equation. For example: if $x = 1$, $y = 6$; if $x = 3$, $y = 4$; if $x = -1$, $y = 8$; if $x = 5$, $y = 2$; etc.

Another equation connecting x and y could be $x - y = 3$. Again there are numerous solutions to this equation such as: if $x = 8$, $y = 5$; if $x = 4$, $y = 1$; etc.

When the two linear equations are considered together then there is only one pair of values that will satisfy both of the equations.

$x + y = 7$

$x - y = 3$

The only pair of values that satisfies both of these equations is $x = 5$, $y = 2$.

The pair of equations $x + y = 7$, $x - y = 3$ are called **simultaneous equations**.

To **solve** a pair of simultaneous equations you have to find values of the unknown letter symbols that work for both equations.

The most popular method is called **the elimination method**.

Example 35.1

Solve the simultaneous equations to find the values of x and y.

$2x + y = 7$

$x - y = 2$

Solution

Step 1
Number the equations.

$$2x + y = 7 \qquad (1)$$
$$x - y = 2 \qquad (2)$$

Step 2
Check to see whether either the numbers in front of x (called the **coefficients of x**) or the numbers in front of y (called the coefficients of y) are the same (ignore the signs at this stage).

In this example you have:

x	y
+2	+1
+1	-1

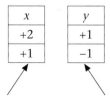

No match:
2 is not the same as 1.

Match:
1 is the same as 1.

> **Reminder**
> Ignore the signs at this stage.

265

Step 3

Add or subtract the equations.

The coefficients of y match but the signs are *different* so you *add* the equations. This will eliminate the y-terms.

$$2x + y = 7 \qquad (1)$$

$$x - y = 2 \qquad (2)$$

Adding equation (1) to equation (2) gives: $\qquad 3x = 9$

Step 4

Dividing both sides by 3 gives: $\qquad x = 3$

Step 5

To find y you now go back to either equation (1) or equation (2) and substitute the value of x.

Using equation (1): $\qquad 2x + y = 7$

Substituting $x = 3$ gives: $\qquad 2(3) + y = 7$

$$6 + y = 7$$

Subtracting 6 from both sides: $\qquad y = 1$

The solution that satisfies both equations is $x = 3$ and $y = 1$.

Step 6

Checking in $x - y = 2$ gives $3 - 1 = 2$.

> **Reminder**
> An equation is like a balance, you have to do exactly the same operation to both sides of the equation to keep it balanced.

> EXAMINER *TIP*
> It does not matter which equation you use to find the y-value.

> EXAMINER *TIP*
> You can check your answers by substituting the numbers back into the other equation.

Example 35.2

Solve the simultaneous equations to find the values of x and y.

$3x + y = 5 \qquad x + y = 3$

Solution

$3x + y = 5 \qquad (1)$

$x + y = 3 \qquad (2)$

The coefficients of y match and the signs are the *same* so you *subtract* the equations. This will eliminate the y-terms.

$$3x + y = 5 \qquad (1)$$

$$x + y = 3 \qquad (2)$$

Subtracting equation (2) from equation (1) gives: $\qquad 2x = 2$

Dividing both sides by 2: $\qquad x = 1$

To find y you now go back to either equation (1) or equation (2) and substitute the value of x.

Using equation (1): $\qquad\qquad\qquad 3x + y = 5$

Substituting $x = 1$ gives: $\qquad\qquad 3(1) + y = 5$

$$3 + y = 5$$

Subtracting 3 from both sides: $\qquad\quad y = 2$

The solution that satisfies both equations is $x = 1$ and $y = 2$.

Checking in $x + y = 3$ gives $1 + 2 = 3$.

Equations with different coefficients

Sometimes the coefficients do not match. These questions will involve one or two extra steps to make the coefficients match.

Example 35.3

Solve the simultaneous equations to find the values of x and y.

$3x + 2y = 10 \qquad x - y = 5$

Solution

$3x + 2y = 10 \qquad\qquad (1)$

$\quad x - y = 5 \qquad\qquad\ (2)$

The coefficients of x do not match and neither do the coefficients of y. In order to proceed you need to multiply equation (2) by 2 so that the y-coefficients will then be the same.

$$3x + 2y = 10 \qquad\qquad (1)$$

Multiplying both sides of $x - y = 5$ by 2: $\qquad 2x - 2y = 10 \qquad\qquad (3)$

In (1) and (3) the coefficients of y match but the signs are *different* so you *add* the equations. This will eliminate the y-terms.

$$3x + 2y = 10 \qquad\qquad (1)$$
$$2x - 2y = 10 \qquad\qquad (3)$$

Adding equation (1) to equation (3) gives: $\qquad 5x = 20$

Dividing both sides by 5: $\qquad\qquad\qquad x = 4$

To find y you now go back to either equation (1) or equation (2) and substitute the value of x.

Using equation (1): $\qquad\qquad 3x + 2y = 10$

Substituting $x = 4$ gives: $\qquad 3(4) + 2y = 10$

$\qquad\qquad\qquad\qquad\qquad\quad 12 + 2y = 10$

Subtracting 12 from both sides: $\qquad 2y = -2$

Dividing both sides by 2: $\qquad\qquad y = -1$

The solution that satisfies both equations is $x = 4$ and $y = -1$.

Checking in $\quad x - y = 5$

gives $\qquad\quad 4 - -1 = 4 + 1$

$\qquad\qquad\qquad\qquad = 5$

> EXAMINER **TIP**
>
> It does not matter which equation you use to find the y-value but always check that your answers work in both of the given equations.

Example 35.4

Solve the simultaneous equations to find the values of x and y.

$5x + 2y = 11 \qquad 2x + 3y = 0$

Solution

$5x + 2y = 11 \qquad\qquad (1)$

$2x + 3y = 0 \qquad\qquad (2)$

In order to proceed you need to multiply equation (1) by 3 and equation (2) by 2 so that the y-coefficients match.

Multiplying both sides of $5x + 2y = 11$ by 3: $\qquad 15x + 6y = 33 \qquad (3)$

Multiplying both sides of $2x + 3y = 0$ by 2: $\qquad 4x + 6y = 0 \qquad (4)$

The coefficients of y match and the signs are the *same* so you *subtract* the equations. This will eliminate the y-terms.

$\qquad\qquad\qquad\qquad\qquad\qquad 15x + 6y = 33 \qquad (3)$

$\qquad\qquad\qquad\qquad\qquad\qquad 4x + 6y = 0 \qquad (4)$

Subtracting equation (4) from equation (3) gives: $\qquad 11x = 33$

Dividing both sides by 11: $\qquad\qquad\qquad\qquad x = 3$

To find y you then go back to either equation (1) or equation (2) and substitute the value of x.

Using equation (1): $\qquad\qquad\qquad 5x + 2y = 11$

Substituting $x = 3$ gives: $\qquad\qquad 5(3) + 2y = 11$

$\qquad\qquad\qquad\qquad\qquad\qquad 15 + 2y = 11$

Subtracting 15 from both sides: $\qquad 2y = -4$

Dividing both sides by 2: $\qquad\qquad y = -2$

The solution that satisfies both equations is $x = 3$ and $y = -2$.

Checking in $\qquad\qquad 2x + 3y = 0$

gives $\qquad\qquad 2 \times 3 + 3 \times -2 = 6 - 6$

$\qquad\qquad\qquad\qquad\qquad = 0$

> *Reminder*
>
> Once the coefficients match then Signs <u>S</u>ame <u>S</u>ubtract, Signs <u>D</u>ifferent A<u>d</u>d.

Practice questions 1

Solve the simultaneous equations

1 $4x + y = 13$ **2** $x + 5y = 1$ **3** $4x + y = 9$
   $x - y = 2$ $2x - y = 13$ $3x + y = 7$

4 $2x + 5y = 8$ **5** $3x - 5y = 11$
   $3x + 4y = 5$ $2x + 3y = 1$

Using simultaneous equations to solve real-life problems

Questions may involve forming a pair of simultaneous equations from a real life-problem, then solving the problem by solving the equations.

Example 35.5

When two numbers are added the answer is 25. The difference between the same two numbers is 9. What are the numbers?

Solution

Let the first number be x and the second number be y.

You are given two statements:

Adding the numbers gives an answer of 25. This means $x + y = 25$.

Subtracting the numbers gives an answer of 9. This means $x - y = 9$.

Rewriting the equations and numbering them gives:

$x + y = 25$ (1)

$x - y = 9$ (2)

You can now go on to solve these simultaneous equations.

The coefficients of y match but the signs are *different* so you *add* the equations. This will eliminate the y-terms.
Adding equation (1) to equation (2) gives: $2x = 34$

Dividing both sides by 2: $x = 17$

To find y you now go back to either equation (1) or equation (2) and substitute the value of x.

Using equation (1): $x + y = 25$

Substituting $x = 17$ gives: $17 + y = 25$

Subtracting 17 from both sides: $y = 8$

So the two numbers are 17 and 8.

Practice questions 2

1   An adult ticket and a child ticket to the cinema costs £6.
    2 adult tickets and 3 child tickets to the cinema cost £14.50.

    Form a pair of simultaneous equations and solve them to find out how
    much an adult ticket costs.

2   5 small bottles of water and 2 large bottles of water hold 6.5 litres
    altogether.
    2 small bottles of water and 5 large bottles of water hold 11 litres
    altogether.
    How much water does each size of bottle hold?

3   In a quiz, Trevor scores twice as many points as Gordon.
    Their total score is 87. What is Gordon's score?

Practice exam questions

1   Solve the simultaneous equations.
    $2x + y = 9$
    $x - 2y = 7$
    You *must* show all your working.   [AQA (SEG) 1998]

2   Solve the simultaneous equations.
    Do not use a trial and improvement method.
    $4x - 3y = 11$
    $6x + 2y = 10$ [AQA (NEAB) 2001]

3   Solve the simultaneous equations.
    $x + 5y = 8$
    $5x - 3y = 5$ [AQA (NEAB) 2001]

4   In a test Adam scored 10 more marks than Rachael.
    Altogether they scored 62 marks.
    Find Adam's score.   [AQA 2003]

5   Solve the simultaneous equations.
    $2x + 3y = 9$
    $3x + 2y = 1$
    You *must* show your working.
    Do *not* use trial and improvement.   [AQA 2003]

6   Solve the simultaneous equations.
    $2x + y = 2$
    $4x - 3y = 9$ [AQA (NEAB) 2000]

36 Solving simultaneous equations by graph

You may be asked to find the solution of a pair of linear equations by using a graphical method.

Simultaneous means 'at the same time', so in order to solve equations graphically you will have to draw both lines on the same axes and read off the solution at the point of intersection.

> **Reminder**
> See Chapter 15 for drawing linear graphs.

Example 36.1

By drawing the graphs of $x + y = 4$ and $y = 2x + 1$ for $-2 \leqslant x \leqslant 5$ find the solution of the simultaneous equations:

$$x + y = 4 \qquad y = 2x + 1$$

Solution

Both graphs are linear (straight lines) so you will need to calculate two points for each.

Consider the graph of $x + y = 4$. This graph passes through the points $(-2, 6)$ and $(5, -1)$.

Consider the graph of $y = 2x + 1$. This graph passes through the points $(-2, -3)$ and $(5, 11)$.

These points and straight lines are plotted on the graph opposite.
The lines on the graph intersect at the point $(1, 3)$ so the solution to the simultaneous equations is $x = 1$, $y = 3$.

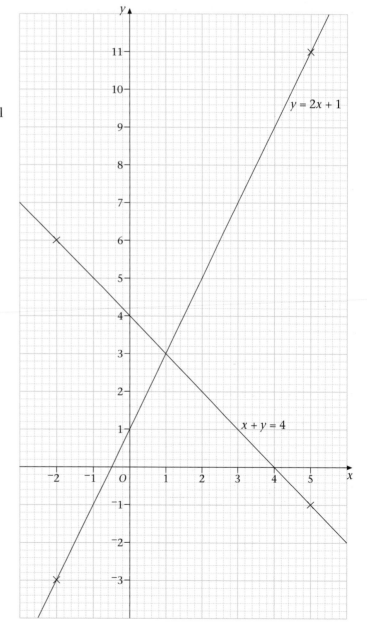

271

Example 36.2

Solve the simultaneous equations $x + 2y = 5$ and $4y = 3x - 3$ by drawing the graphs on the grid.

Solution

Set up a table of coordinates for each graph.

$x + 2y = 5$

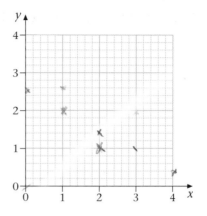

x	0	1	2	3	4
y	2.5	2	1.5	1	0.5

$4y = 3x - 3$

x	0	1	2	3	4
y	−0.75	0	0.75	1.5	2.25

Plot the points accurately and find the point of intersection of the two graphs.

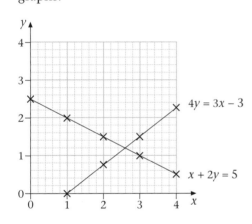

The solution to this pair of simultaneous equations is $x = 2.6$, $y = 1.2$.

Practice question

1   By drawing the graphs, find the solution of each pair of simultaneous equations.

a $y = 4x + 1$
$y = -2$

b $y = -x$
$y = 2x - 3$

c $x + y = 5$
$x - y = 1$

d $y = x + 4$
$y = 2x + 1$

e $y = 3x - 1$
$x + 3y = 2$

f $y = 2x + 5$
$y = -3x$

g $3x + 2y = 7$
$x - 2y = 5$

Practice exam questions

1 The line $y = \frac{1}{2}x + 1$ is drawn on the grid below.

  a Write down the gradient of the line $y = \frac{1}{2}x + 1$.

  b Copy the graph and, on the same grid, draw the graph of $y = 6 - 2x$.

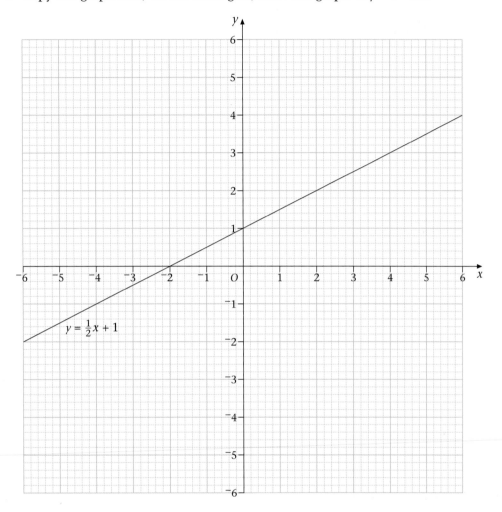

  c Use your graphs to solve the simultaneous equations $y = \frac{1}{2}x + 1$ and $y = 6 - 2x$.   [AQA (NEAB) 2001]

2 Use a graphical method to solve the simultaneous equations $y = x + 7$ and $y + 3x = 5$.

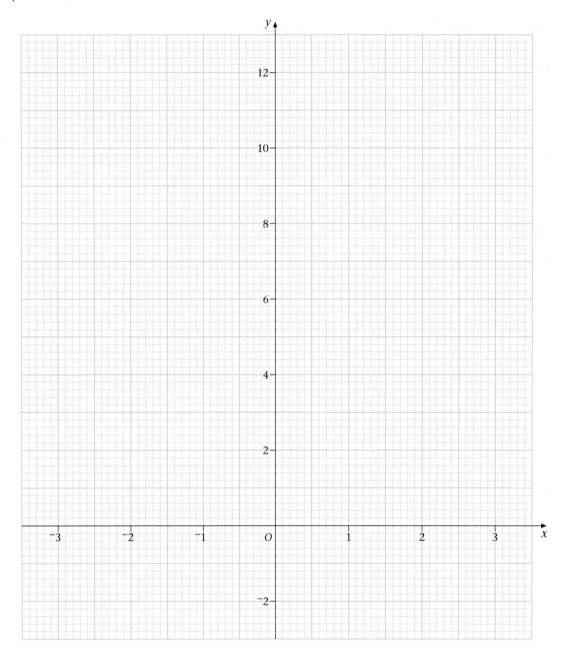

[AQA (NEAB) 2000]

37 Volume

The **volume** of a three-dimensional shape is the amount of space that it occupies.

A cube which measures 1 centimetre long, 1 centimetre wide and 1 centimetre high has a volume of 1 **cubic centimetre**.

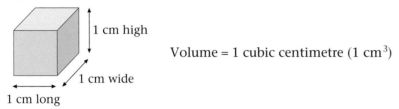

Volume = 1 cubic centimetre (1 cm^3)

Volumes are measured in cubic units.

You write 1 cubic centimetre as 1 cm^3.

Small volumes may be measured in cubic millimetres (mm^3).

Large volumes may be measured in cubic metres (m^3).

Volume of a cuboid

A cuboid is a rectangular box.

If you cut the cuboid at right angles to its length you can see that the cuboid will have the same cross-section throughout its length. This shape is said to have a **uniform cross-section** because its cross-section is the same shape all the way through the length of the object.

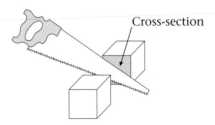

Cross-section

A cuboid can have a square cross-section or a rectangular cross-section.

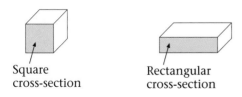

Square cross-section

Rectangular cross-section

A simple cuboid could be made by placing eight centimetre cubes side by side in a line.

The volume of this cuboid is 8 cm³.

The volume of a cuboid is the amount of space it occupies. You can think of this as how many centimetre cubes will fit inside the cuboid.

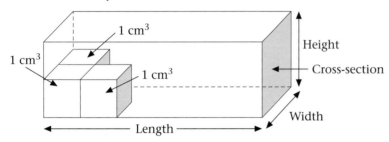

The formula for the volume of a cuboid is:

Volume = length × width × height

$V = l \times w \times h$

If this cuboid is 6 cm long, 2 cm wide and 2 cm high then 6 × 2 × 2 = 24 centimetre cubes will fit exactly inside the cuboid. The volume of this cuboid will be 24 cm³.

EXAMINER **TIP**

'Width' is sometimes referred to as 'breadth' so you may see the formula written as $V = l \times b \times h$. The formula can be written without the multiplication signs as $V = lwh$ or $V = lbh$.

Example 37.1

Find the volume of each cuboid.

Cuboid	Length	Width	Height
a	5 cm	2 cm	4 cm
b	3 cm	6 cm	3 cm
c	8 mm	12 mm	5 mm

Solution

Volume = length × width × height

$V = l \times w \times h$

a $V = 5 \times 2 \times 4$
    $= 40$ cm³

b $V = 3 \times 6 \times 3$
    $= 54$ cm³

c $V = 8 \times 12 \times 5$
    $= 480$ mm³

EXAMINER **TIP**

Make sure you are working in the correct units.

Example 37.2

A cuboid has sides measuring 32 mm by 50 mm by 25 cm.

Calculate the volume of this cuboid.

Solution

Drawing a diagram will often help you to visualise the shape and make the problem easier to understand.

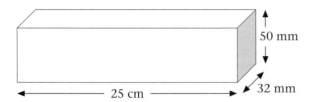

The lengths of the sides are given in different units. Two sides are given in millimetres and the length is in centimetres.

You can only calculate volumes when all of the measurements are in the same units.

You can change these units either all into millimetres or all into centimetres.

> **Reminder**
> 10 millimetres = 1 centimetre

Method A (answer in cubic millimetres)

$1 \text{ cm} = 10 \text{ mm}$
$25 \text{ cm} = 250 \text{ mm}$

$$V = lwh$$
$$= 250 \times 32 \times 50$$
$$= 400\,000 \text{ mm}^3$$

Method B (answer in cubic centimetres)

$10 \text{ mm} = 1 \text{ cm}$
$32 \text{ mm} = 3.2 \text{ cm}$
$50 \text{ mm} = 5 \text{ cm}$

$$V = lwh$$
$$= 25 \times 3.2 \times 5$$
$$= 400 \text{ cm}^3$$

> **EXAMINER TIP**
> Be careful with units as it is easy to get confused when changing from one unit into another.

Both answers are fully correct. This means that $400\,000 \text{ mm}^3 = 400 \text{ cm}^3$.

So $1000 \text{ mm}^3 = 1 \text{ cm}^3$.

Volume of a cube

A cube has length = width = height.

This cube has sides of length x units.

The volume of the cube is $V = x \times x \times x$ or $V = x^3$.

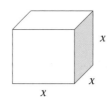

Example 37.3

Calculate the volume of each cube.

a   side = 5 cm   b   side = 10 cm   c   side = 2 m   d   side = 1.5 mm

Solution

a $V = 5 \times 5 \times 5$ b $V = 10 \times 10 \times 10$ c $V = 2^3$ d $V = (1.5)^3$
    $= 125 \text{ cm}^3$ $= 1000 \text{ cm}^3$ $= 8 \text{ m}^3$ $= 3.375 \text{ mm}^3$

Practice questions 1

1   Calculate the volume of each of the following cuboids.

   a   length = 24 cm, width = 6 cm, height = 7.5 cm
   b   length = 3.7 cm, width = 5.8 cm, height = 4.2 cm
   c   length = 2.4 m, width = 60 cm, height = 28 cm
   d   length = 35 mm, width = 54 mm, height = 18 mm
   e   length = 84 cm, width = 36 cm, height = 75 cm
   f   length = 8.5 cm, width = 44 mm, height = 1.2 cm

2   A water tank is a cuboid with length 1.3 m, breadth 0.75 m and height 85 cm.
   Calculate how much water the tank will hold when full.
   Give your answer in litres.

> **Reminder**
> 1000 cm^3 = 1 litre

3   A cube has each side 10 mm long.

   a   Calculate the volume of the cube. Give your answer in cubic millimetres.
   b   Calculate the volume of the cube. Give your answer in cubic centimetres.
   c   How many cubic millimetres are there in a cubic centimetre?

Prisms

A **prism** is a three-dimensional (3D) object which has a uniform (constant) cross-section.

In the exam you will only be asked questions about right prisms. These are where the end of the prism, the cross-section, is perpendicular to the length of the prism.

The shape of the cross-section gives the name of the prism, e.g. triangular prism or hexagonal prism.

These shapes are all prisms.

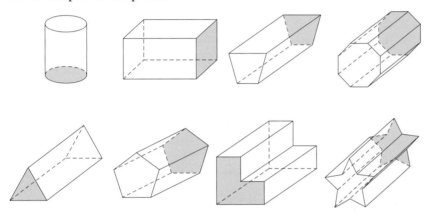

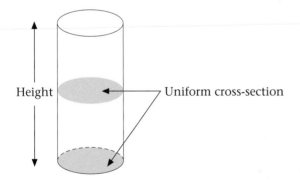

The volume of any prism can be found by multiplying the area of the cross-section by the length (height) of the prism.
A cylinder has a circular end which is the same shape throughout the length of the cylinder, so it has a uniform cross-section.

Height   Uniform cross-section

The volume of the cylinder is the area of the cross-section (the circle) multiplied by its height.

V = area of the cross-section × height

$\quad = \pi r^2 \times h$

$\quad = \pi r^2 h$

Example 37.4

Calculate the volume of the following cylinders.

a radius = 5 cm, height = 30 cm   **b** diameter = 1 m, length = 15 m
c radius = 1.6 cm, length = 25 m

Solution

a $V = \pi r^2 h$
$\quad = \pi \times 5^2 \times 30$
$\quad = 750\pi \text{ cm}^3$
$\quad = 2356.2 \text{ cm}^3$

b $V = \pi r^2 h$
The diameter is 1 m so the radius is
0.5 m.
$\quad V = \pi \times (0.5)^2 \times 15$
$\quad\quad = 3.75\pi \text{ m}^3$
$\quad\quad = 11.78 \text{ m}^3$

c $V = \pi r^2 h$
The length is in metres: 25 m = 2500 cm
$V = \pi \times (1.6)^2 \times 2500$
$\quad = 6400\pi \text{ cm}^3$
$\quad = 20\ 106.2 \text{ cm}^3$

Example 37.5

A triangular prism has base 8 cm, perpendicular height 6 cm and length
30 cm.

Calculate the volume of this prism.

Solution

A large diagram showing this information will help.

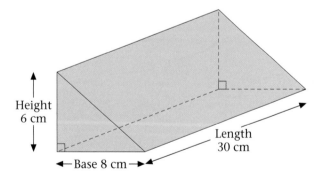

Volume of a prism is given by:

V = area of the cross-section × length
$\quad$ = area of triangle × length
$\quad = \frac{1}{2}bh \times l$
$\quad = \frac{1}{2} \times 8 \times 6 \times 30$
$\quad = 720 \text{ cm}^3$

> **Reminder**
> Area of a triangle is $\frac{1}{2}bh$.

 EXAMINER *TIP*
Check that you are using the
perpendicular height of the
triangle to calculate its area.

Example 37.6

A metal girder has a uniform cross-section in the shape of an L as shown. The girder is 2 metres long.

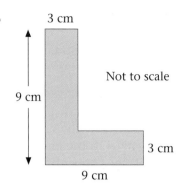

a Calculate the volume of the girder.

b The density of the metal is 1.5 grams per cm^3. Calculate the weight of the girder, giving your answer in kilograms.

Solution

a Volume of girder = area of cross-section × length
  Area of cross-section = $(3 \times 6) + (3 \times 9)$
  $= 18 + 27$ cm^2
  $= 45$ cm^2
  Length = 2 m
  = 200 cm (changing units from metres to centimetres)

  Then the volume of the girder is $45 \times 200 = 9000$ cm^3.

b The density of the metal is 1.5 grams per cm^3.
  This means that every 1 cm^3 of the metal girder weighs 1.5 grams.
  So 9000 cm^3 will weigh 9000×1.5 g.

  Weight of girder = 9000×1.5 g
  = 13 500 g

  So, in more appropriate units, the girder weighs 13.5 kg.

Practice questions 2

1 Calculate the volumes of the following prisms.

  a cylinder with radius 12 cm and height 2.5 m
  b triangular prism with base 10 cm, perpendicular height 5 cm and length 34 cm
  c prism with a semicircular cross-section of radius 8 mm and length 65 mm
  d diamond prism with sides 5 cm and length 24 cm

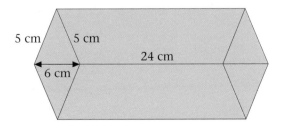

2 A copper pipe is 0.1 cm thick and is 4 m long.
The cross-section of the pipe is shown.
The inner radius of the pipe is 0.9 cm.

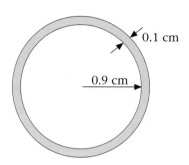

0.1 cm

0.9 cm

   a Calculate the area of the shaded cross-section of the pipe.

   b Calculate the volume of the copper metal in the pipe.

   c If the copper has a density of 9 g per cm^3 calculate the weight of this pipe.

Volume of a cone

Cones and pyramids do not have a uniform cross-section and their volumes are calculated differently to prisms.

The volume of any pyramid is
$\frac{1}{3}$ × base area × perpendicular height.
A **cone** is a pyramid with a circular base.

The volume of a cone is given by $V = \frac{1}{3}\pi r^2 h$.

You may notice that the volume of a cylinder is $\pi r^2 h$ and the volume of a cone is $\frac{1}{3}\pi r^2 h$.

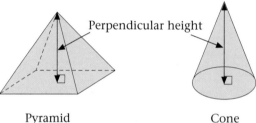

Perpendicular height

Pyramid   Cone

This means that the volume of three cones is the same as the volume of a cylinder which has the same radius and height as the cones.

Note: You will be given the formula for the volume of a cone in the examination if it is needed.

Example 37.7

Calculate the volume of a cone with radius 4 cm and perpendicular height 15 cm.

Solution

The volume of the cone is given by:
$V = \frac{1}{3}\pi r^2 h$
$ = \frac{1}{3} \times \pi \times 4^2 \times 15$
$ = 80\pi \text{ cm}^3$
$ = 251 \text{ cm}^3$

Practice question 3

1   Calculate the volume of each cone.

   a radius = 5 cm, perpendicular height = 10 cm
   b radius = 8 cm, perpendicular height = 8 cm
   c radius = 25 cm, perpendicular height = 60 cm
   d radius = 2 m, perpendicular height = 5 m

Practice exam questions

1   A solid plastic cuboid has dimensions 3 cm by 5 cm by 9 cm.
   The density of the plastic is 0.95 grams per cm^3.

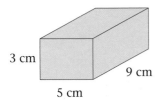

3 cm
9 cm
5 cm

   What is the weight of the plastic cuboid?   [AQA (NEAB) 1998]

2   A water-trough has a semicircular cross-section, as shown.

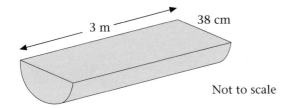

3 m   38 cm

Not to scale

   The diameter of the end of the trough is 38 cm.
   The trough is 3 m long.
   Calculate the volume of water in the trough when it is full.
   Give your answer in litres.   [AQA (SEG) 2002]

3   The volume of a cube is given by the formula $V = L^3$.

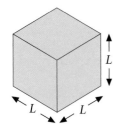

L
L L

   a A cube has volume 5832 cm^3.
        What is the length of each edge?
   b The area of each face of a different cube is 625 cm^2.
        What is the volume of this cube?   [AQA (NEAB) 1999]

4 A cylinder of radius 7 cm and height 18 cm is half full of water.
One litre of water is added.
Will the water overflow?
You must show all your working.

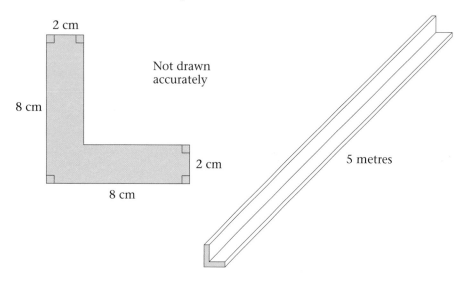

[AQA (NEAB) 2001]

5 A girder is 5 metres long.
Its cross-section is L-shaped as shown below.

2 cm

Not drawn
accurately

8 cm

2 cm

8 cm

5 metres

Find the volume of the girder.
Remember to state the units in your answer.

[AQA (NEAB) 2000]

6 **a** Calculate the area of a circle of diameter 12 cm.

The cross-section of a lead drainpipe is shown in the diagram.
The lead is 0.3 cm thick.

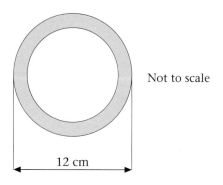

Not to scale

12 cm

b Calculate the area of the shaded cross-section.
c A section of this drainpipe, which is 60 cm long, weighs 7900 g.
Calculate the density of the lead.

[AQA (SEG) 2000]

38 Nets and surface areas

Nets

A **net** is a two-dimensional diagram that can be cut out and folded to make a three-dimensional shape.

More than one net can make the same shape. Two different nets for a cube are shown.

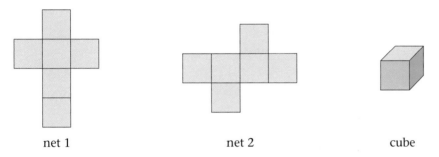

net 1   net 2   cube

Here are some more nets.

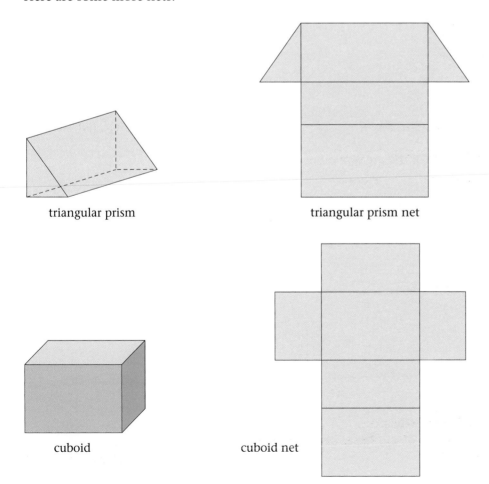

triangular prism   triangular prism net

cuboid   cuboid net

Practice questions 1

1   Sketch a net of a square based pyramid.

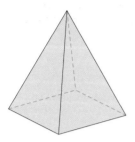

2   Sketch the net of an *open-ended* cylinder.

3   What 3D object would the following net form?

Surface areas

The surface area of a 3D shape is the sum of the areas of all of the faces of the shape.

It is useful to sketch a net of the 3D shape so that you can calculate the area of each face and then sum them to find the total surface area.

Example 38.1

Calculate the surface area of a cube of side 10 cm.

Solution

Sketch a net of a cube.
There are 6 square faces.

The area of each face is $10 \times 10 = 100$ cm^2.

The total surface area is then $6 \times 100 = 600$ cm^2.

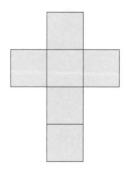

Reminder
Don't forget to state the units in your answer.

Example 38.2

Calculate the total surface area of a cuboid measuring
5 cm × 8 cm × 15 cm.

EXAMINER **TIP**

◄— 5 cm × 8 cm × 15 cm means
that the length is 15 cm, the
height is 8 cm and the width is
5 cm.

Solution

Draw a sketch of the cuboid and its net.

You do not need to make an accurate drawing.

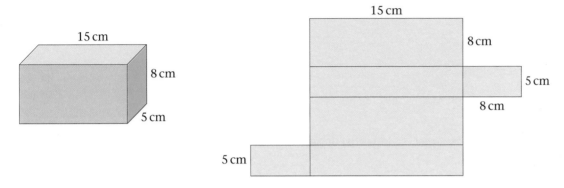

From the net there are:

2 large rectangles (front and back) measuring 15 cm × 8 cm

2 rectangles (top and bottom) measuring 15 cm × 5 cm

2 small rectangles (the two ends) measuring 5 cm × 8 cm.

The areas are:

area of front face is $15 \times 8 = 120$ cm^2

area of back face is $15 \times 8 = 120$ cm^2

area of top face is $15 \times 5 = 75$ cm^2

area of bottom face is $15 \times 5 = 75$ cm^2

area of left end face is $5 \times 8 = 40$ cm^2

area of right end face is $5 \times 8 = 40$ cm^2

Add all these areas together.

The total surface area of this cuboid is 470 cm^2.

Example 38.3

Calculate the *curved surface area* of a cylinder of radius 3 cm and height 15 cm.

Solution

Sketch an open-ended cylinder and its net.

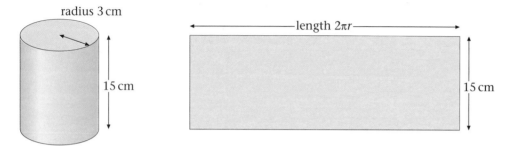

The net of an open-ended cylinder is a rectangle whose length is the circumference of the end circle.

Imagine the rectangle being wrapped around to form a cylinder, this is called the curved surface area.

The length of the rectangle is $2\pi r = 2 \times \pi \times 3$
$$= 18.85 \text{ cm.}$$

The height of the rectangle is 15 cm.

So the area of the rectangle is $18.85 \times 15 = 282.74 \text{ cm}^2$.

Then the curved surface area of this cylinder is 282.74 cm^2.

Practice questions 2

1 Find the total surface area of the following cubes.

   a side 5 cm   **b** side 8 cm   **c** side 15 cm
   d side 1.2 m   **e** side 24 mm

2 Calculate the total surface area of the following cuboids.

Cuboid	length	width	height
a	14 cm	6 cm	10 cm
b	25 mm	10 mm	7 mm
c	1.4 m	0.75 m	0.9 m
d	59 cm	24 cm	3 cm

3 Find the curved surface areas of the following cylinders.

   a radius 10 cm, height 25 cm
   b radius 23 mm, length 65 mm
   c radius 30 cm, length 4 m

4 Calculate the total surface areas of the cylinders in question 3, assuming the cylinders are closed cylinders.

5 Calculate the total surface area of the triangular prism below.

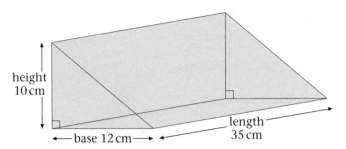

height
10 cm

←— base 12 cm —→

length
35 cm

Practice exam questions

1 Nets are to be made using different sizes of the following shapes.

Square   Triangle   Rectangle

Example: a cuboid

2 squares, 0 triangles, 4 rectangles

Give the number of each shape required to complete a net of:

a a cube
b a square based pyramid
c a triangular prism.

[AQA (NEAB) 2000]

2 The diagram shows a model tower.
It is a square based pyramid on top of a cuboid.
  a How many planes of symmetry has the tower?
  b Sketch a net of this tower including the base.
  You do *not* need to make an accurate drawing.

Reminder
For help with planes of
symmetry see chapter 16.

[AQA (NEAB) 2002]

3 **a** From the list below, write down the correct name for each of the solid
  shapes.

  cylinder
  sphere
  prism
  cone
  pyramid

  b Four students have each drawn a net for a cube.

Hannah	Jared	Kirsty	Liam
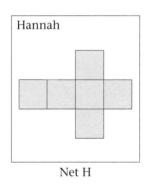	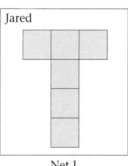	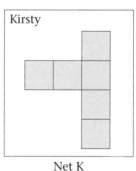	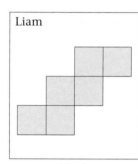
Net H	Net J	Net K	Net L

  Which one of these nets would *not* fold up to make a cube?   [AQA (SEG) 2000]

4 A triangular prism has dimensions as shown.
  a *Sketch* a net of the prism.
  (You do *not* need to draw an accurate diagram.)
  b Calculate the total surface area of the prism.
  Show all your working.

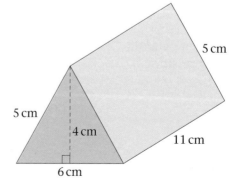

[AQA (NEAB) 2001]

5 The diagram shows a net of a cube.
The side of the cube is 5 cm.

  a Calculate the area of the net,
    stating your units.

The net is folded to make the cube.

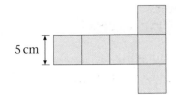

  b Calculate the volume of the cube.

[AQA (SEG) 2000]

6 The diagram shows a triangular prism.

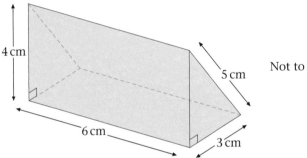

Not to scale

Part of the net of this prism has been drawn below.

  a Complete the accurate drawing of the net.

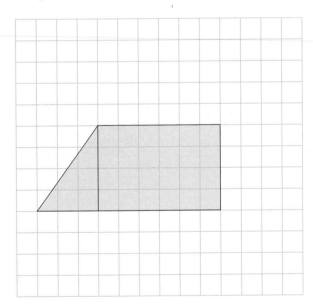

  b Calculate the total surface area of the prism.
    State your units.

[AQA (SEG) 2002]

39 Circle theorems

You will need to know and use the following facts about angles in circles and tangents to circles.

Angle in a semicircle

Any angle at the circumference of a circle drawn from the diameter, as shown, is a right angle.

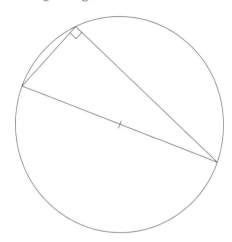

Example 39.1

Work out the value of angle *x*.

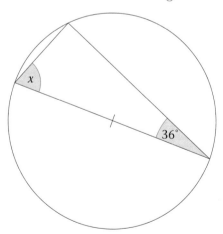

Not drawn accurately

Solution

You are expected to realise that the third angle in the triangle is 90° as it will not be marked on the diagram.

The sum of the angles in a triangle is 180°, so:

$x = 180° - 90° - 36°$
$\quad = 54°$

Angles on the same arc

Any angles on the same arc at the circumference of a circle are equal.

In the diagram $x = y$.

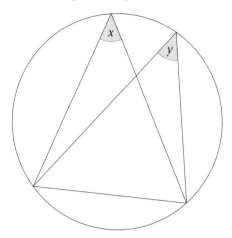

Example 39.2

Work out the value of angle x.

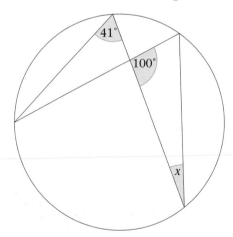

Solution

Fill in the other 41° angle on the diagram:

Now use the fact that sum of the angles in the triangle is equal to 180° to work out the value of x:

$x = 180° - 100° - 41°$
$\quad = 39°$

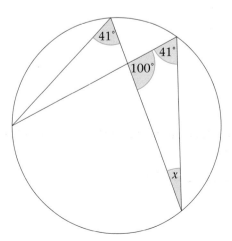

Angle at the centre of the circle and angle at the circumference of the circle

The angle at the centre of the circle is twice the angle at the circumference. For this to be true, both angles must be formed from the same chord (or same two points). They must also be in the same segment.

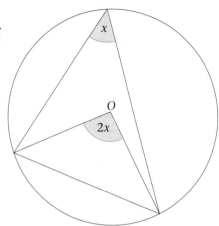

Example 39.3

Work out the value of angle x.

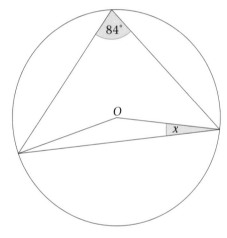

Not drawn accurately

Solution

Fill in the angle at the centre of the circle ($2 \times 84° = 168°$):

You now have an isosceles triangle because two sides are radii of the circle.

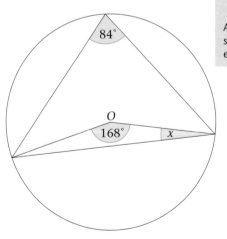

Not drawn accurately

So the unmarked angle is also x.

Therefore $2x + 168° = 180°$.

Subtracting 168° gives:

$2x = 12°$
$x = 6°$

Opposite angles of a cyclic quadrilateral

A **cyclic quadrilateral** is a quadrilateral drawn inside a circle with all four vertices on the circumference.

The **opposite angles** of a cyclic quadrilateral always add up to 180°.

In the diagram $x + y = 180°$.

> **Reminder**
> A quadrilateral is a polygon with four sides and four vertices (corners).

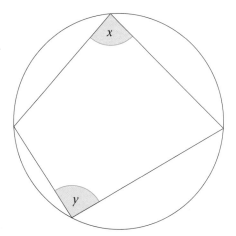

Example 39.4

O is the centre of the circle.

Work out the value of angle *x*.

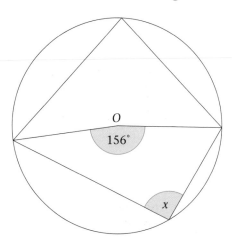

Not drawn accurately

Solution

The angle at the centre of the circle is twice the angle at the circumference.

Filling in the angle at the circumference (156° ÷ 2 = 78°) gives:

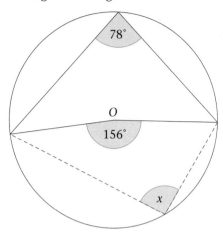

The angles which are x and 78° are opposite angles of the cyclic quadrilateral.

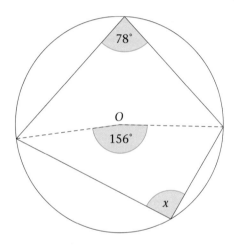

This gives $x + 78° = 180°$

So $x = 102°$

Angle between a tangent to a circle and the radius at the same point

The angle between the tangent to a circle and the radius at the same point is always 90°.

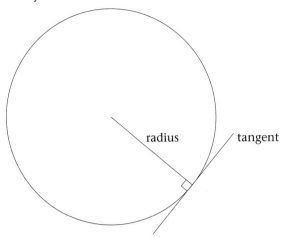

Example 39.5

The diagram shows a tangent to a circle. *O* is the centre of the circle.

Work out the value of angle *x*.

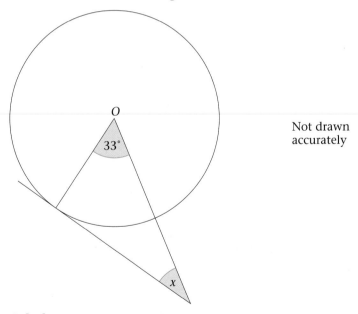

Not drawn accurately

Solution

The angle between the tangent and the radius is 90°.

You can then use the fact that the sum of the angles in the triangle is equal to 180° to work out the value of *x*.

This gives $x = 180° - 90° - 33°$

So $x = 57°$

Tangents to a circle from a point outside the circle

It is always possible to draw two tangents to a circle from a fixed point outside the circle. The distances of the fixed point to the points of contact with the circle are equal.

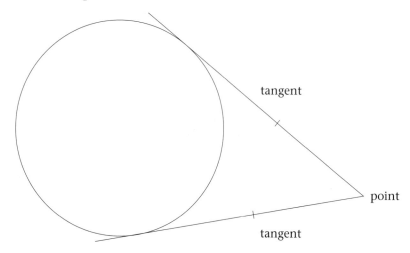

If you now draw a line to the centre of the circle and then put in the radii to the points of contact of the tangents, you will have two congruent triangles (see Chapter 34).

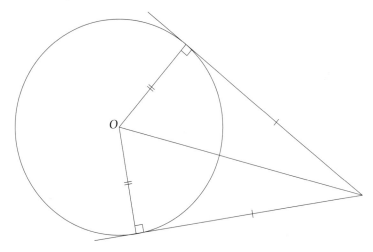

Questions could then involve the use of Pythagoras' theorem (see Chapter 23) or trigonometry (see Chapter 24).

Example 39.6

A circle of radius 5 cm is shown. *O* is the centre of the circle.

Calculate the value of *x*, the length of a tangent to the circle.

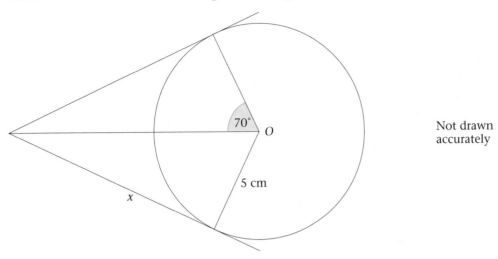

Not drawn
accurately

Solution

As you have two tangents and two radii, the diagram includes two congruent right-angled triangles.

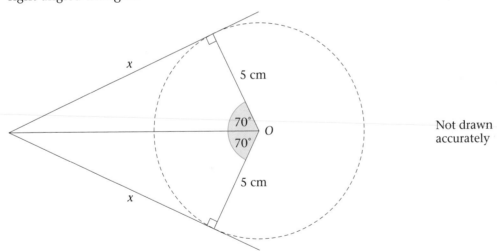

Not drawn
accurately

This is now a question about trigonometry in a right-angled triangle using the tan ratio.

$$\tan 70° = \frac{\text{opposite}}{\text{adjacent}}$$

$$= \frac{x}{5}$$

Multiplying through by 5 gives:

$$5 \times \tan 70° = x$$

$$13.7 \text{ cm} = x$$

Example 39.7

Explain why the perpendicular from the centre of a circle to a chord bisects the chord.

Reminder
Perpendicular means 'at right angles to'.

Solution

Looking at the diagram, the perpendicular means that there is a right angle in each of the two small triangles.

The radius is also in both small triangles.

The perpendicular forms a common side.

Reminder
Bisects means 'cuts into two equal parts'.

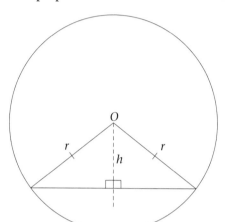

So the third sides in the small triangles are equal.

This means that the small triangles are congruent.

This means that the perpendicular from the centre of the circle to the chord bisects the chord.

Reminder
See Chapter 34 for help on congruency.

Practice questions

In each question *O* is the centre of the circle.

1   Find the value of the angle marked with a letter in each part, giving a reason for your answer.

a

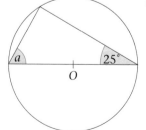

b

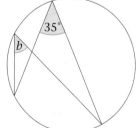

c

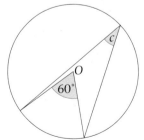

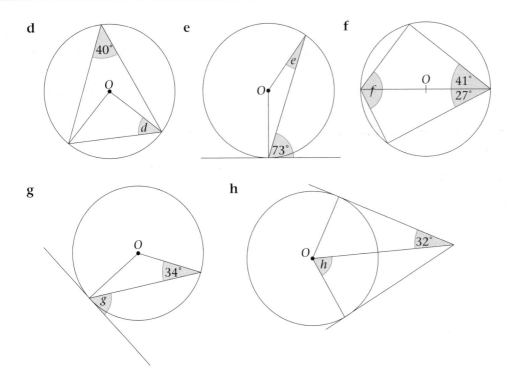

d   **e**   **f**

g   **h**

2 Work out:

a angle *BAC*   **b** angle *BAC*   **c** angle *AOC*

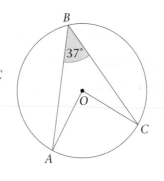

d angle *ABC*   **e** angle *OST*   **f** angle *ABC*

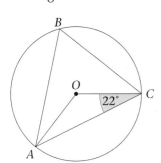   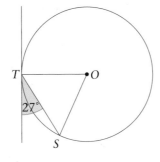   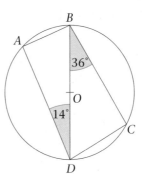

g angle *TPQ*   **h** angle *ABC*

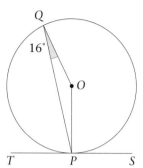 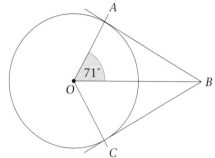

Practice exam questions

1 In the diagram, *O* is the centre of a circle and *P, Q, R, S* are points on the circumference.
Angle *POR* = 112°

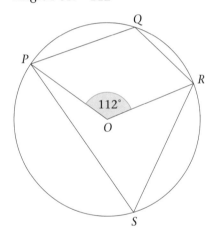

Not to scale

   a Calculate the size of angle *PSR*, giving a reason for your answer.
   b Calculate the size of angle *PQR*, giving a reason for your answer.   [AQA (SEG) 2000]

2 *K, L, M, N* are four points
on the circumference of a
circle.
Angle *LNM* = 70° and
angle *KML* = 24°
Calculate the following
angles, giving reasons for
your answers.

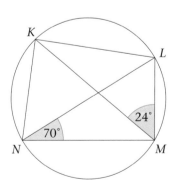

Not to scale

   a angle *KNL*
   b angle *KLM*   [AQA (SEG) 2001]

3   In the diagram below, *A*, *B*, *C* and *D* lie on a circle.
    BAE and *CDE* are straight lines.

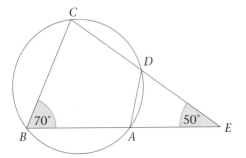

Not to scale

Work out the size of angle *DAE*.
Give a reason for each step of your working.

[AQA (SEG) 2000]

4   *A*, *B* and *C* are points on the circumference of a circle centre *O*.
    CT is the tangent to the circle at *C*.
    Angle *AOC* = 116°
    Angle *BCT* = 64°

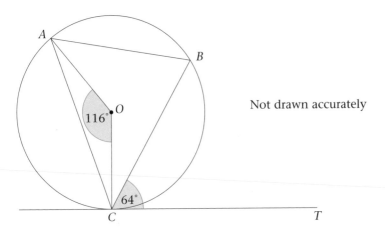

Not drawn accurately

a   Write down the size of angle *ABC*.
b   Show that triangle *ABC* is isosceles.
    Show your working clearly.
    Give reasons for any values calculated.

[AQA (NEAB) 2001]

40 Proof

Proof is a topic that will require you to be able to give clear explanations, show how statements are true or false using examples or apply mathematical reasoning to a solution.

Proof questions can be set on a variety of topics including number, algebra and geometry, but will always be based on your knowledge from other sections of the module.

Questions will be set at various levels of difficulty.

Counter examples

You may be asked to give examples to show that a statement is *not true*. This is called a **counter example**.

Example 40.1

Jonathan says:

TO ADD TWO FRACTIONS YOU JUST ADD THE TOPS AND ADD THE BOTTOMS.

Give an example to show the correct method of adding fractions. Show that Jonathan's rule does not work.

Solution

To add fractions you should use the common denominator method.

For example, looking at $\frac{2}{3} + \frac{1}{2}$, the common denominator is 6.

So $\frac{2}{3} + \frac{1}{2} = \frac{4}{6} + \frac{3}{6}$

$$= \frac{7}{6}$$

$$= 1\frac{1}{6}$$

> **Reminder**
> See Chapter 2 for work on equivalent fractions.

Using the incorrect rule would give an answer of $\frac{3}{5}$ which clearly does not work.

Proving an identity

You may be asked to prove that one algebraic statement is equivalent to another. This could involve factorising or expanding brackets.

Example 40.2

Show that $(x - 3)^2 = x^2 - 6x + 9$.

Solution

When you are asked to show that an algebraic statement is true, it is important that you show every step of working whatever method you decide to use.

$(x - 3)^2 = (x - 3)(x - 3)$

$\qquad = x^2 - 3x - 3x + 9$

$\qquad = x^2 - 6x + 9$

> *Reminder*
> See Chapter 5 for work on expanding brackets.

Proving the properties of a geometrical shape

You may be asked to prove a piece of information about the properties of a geometrical shape. This could involve your knowledge of angle facts.

Example 40.3

In the diagram triangle *ABC* is similar to triangle *ADE*.

Prove that *BC* is parallel to *DE*.

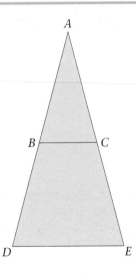

Solution

Angle *A* is common to both triangles.

As triangle *ABC* is similar to triangle *ADE* this means that

Angle *ABC* = Angle *ADE*.

So you have corresponding angles (F-angles) which means that

BC is parallel to *DE*.

> *Reminder*
> See Chapter 17 for work on parallel lines.

Practice questions

1   **a**   Show that $(x + 2)^2 = x^2 + 4x + 4$.

    b   Simplify $\dfrac{x + 4x + 4}{x^2 + 2x}$.

2   Two congruent right-angled triangles are pieced together.
    Show how you can make:

    a   an isosceles triangle
    b   a parallelogram
    c   a rectangle.

3   An isosceles triangle *ABC* is shown.
    $AB = AC$
    Angle $A = 2x$
    Show that the exterior angle at *B* is $90° + x$.

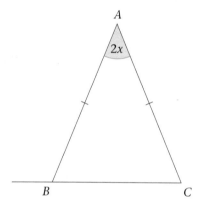

4   Prove that the angle sum of a triangle is 180°.

5   Prove that an exterior angle of a triangle is equal to the sum of the interior
    opposite angles.

Practice exam questions

1   The length of a rectangle is twice its width.

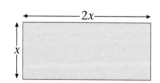

Not to scale

    The area of the rectangle is 50 cm^2.

    a   Show that the length of the rectangle is 10 cm.
    b   Work out the perimeter of the rectangle.

    [AQA 2003]

2 A shape is made up of two squares and an equilateral triangle as shown.

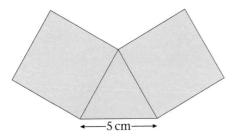

←—5 cm—→

Rebecca says,

The area of the shape is
greater than 50 cm^2
and less than 75 cm^2

Is she correct?
Give a reason for your answer.

[AQA (NEAB)

Practice exam paper

Formula sheet

Formulae Sheet: Intermediate Tier

You may need to use the following formulae:

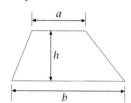

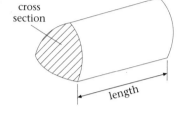

Area of trapezium $= \frac{1}{2}(a + b)h$

Area of prism = area of cross section × length

Paper 1   Non-calculator   1 hour 15 minutes

Total of 70 marks for Paper 1

1 a Work out the value of 2^5.

....

Answer .... *(1 mark)*

b Write down the value of $\sqrt[3]{1000}$.

....

Answer .... *(1 mark)*

2 a Find the value of $2x - y$ when $x = 7$ and $y = -1$.

....

Answer .... *(2 marks)*

b Simplify $a + 4b + 8a - 6b$.

....

Answer .... *(2 marks)*

3 The graph shows the volume of water in a leaking barrel.

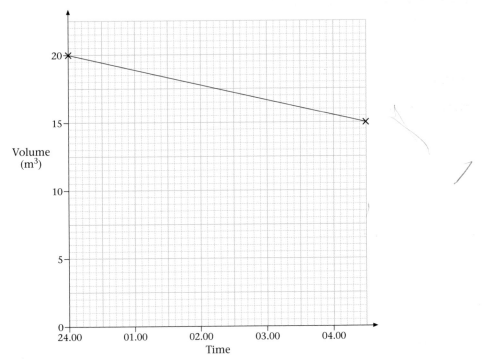

The water continues to leak at the same rate until the barrel is empty.
At what time will the barrel be empty?

....

....

....

Answer.. *(3 marks)*

4 The perimeter of a regular hexagon is 42 cm.
The hexagon is cut up to make six equilateral triangles.

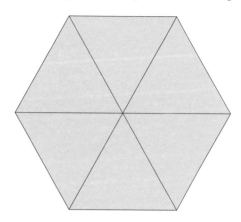

Work out the perimeter of one triangle.

....

....

Answer.. *(2 marks)*

5 The diagram shows the position of *A*. The diagram is drawn to scale.

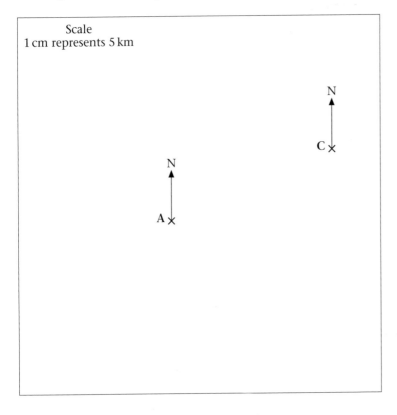

a *B* is 20 km from *A* on a bearing of 090°.
Mark the position of *B* on the diagram.

....

.... *(2 marks)*

b Write down the bearing of *A* from *B*.

....

Answer .... *(1 mark)*

c How far is the actual distance of *C* from *A*?

....

Answer .... *(2 marks)*

6

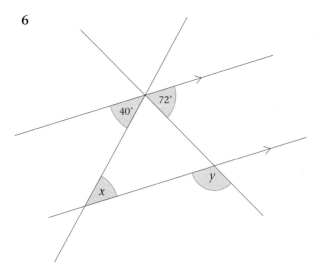

a Work out the value of *x*.

....

Answer .... *(1 mark)*

b Work out the value of *y*.

....

....

Answer .... *(2 marks)*

7 The diagram shows a triangle with base = 16 cm and perpendicular height = 4 cm.

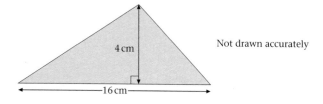

Calculate the area of the triangle.

....

....

....

Answer .... *(3 marks)*

8 a The *n*th term of a sequence is $3n + 2$.

    i Write down the first **three** terms of the sequence.

....

                  Answer.. *(2 marks)*

    ii Explain why 30 is not a term in the sequence.

....

.... *(1 mark)*

b Write down the *n*th term of this sequence:

    4   7   10   13

....

                  Answer .... *(1 mark)*

9 Matt is 3 years older than Dan. The sum of their ages is 27 years.
How old is Matt?

....

....

....

                  Answer.. *(3 marks)*

10 a The diagram shows a shaded shape *A*.

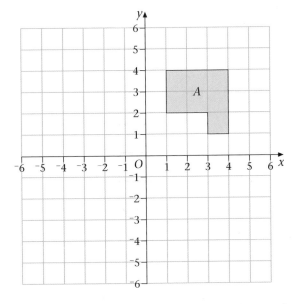

    Rotate the shape *A* through 90° clockwise about *O*.   *(3 marks)*

b The diagram shows three shapes *A*, *B* and *C*.

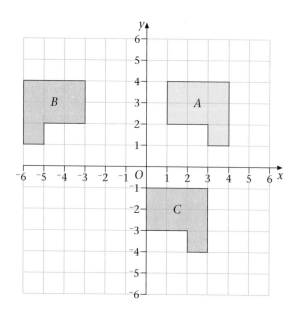

i Describe fully the **single** transformation which maps shape *A* to shape *B*.

....

....

Answer... *(2 marks)*

ii Shape *A* is translated to shape *C*.
Write down the vector that describes the translation.

Answer... *(2 marks)*

11 The cross-section of a shed is a pentagon.
Sketch the plan view of the shed.

(2 marks)

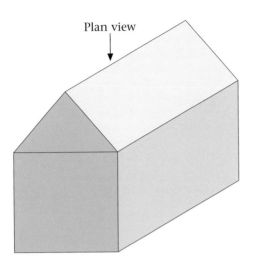

Plan view

12 The exterior angle of a regular polygon is 36°.

Not drawn accurately

36°

a Work out the total number of sides of the polygon.

....

....

Answer.. (2 marks)

b O is the centre of the regular polygon.

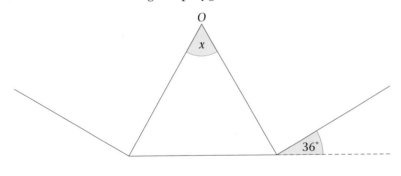

O

x

36°

Not drawn accurately

Work out the size of the angle *x*.

....

....

....

Answer.. (3 marks)

13 a Expand and simplify

$3(5x + 4) - 2(x - 1)$

....

....

Answer... *(2 marks)*

b Expand

$x(x^2 + 4)$

....

....

Answer... *(2 marks)*

14 a Complete the table for $y = \dfrac{12}{x}$.

x	1	2	3	4	6	12
y	12	6				1

(2 marks)

b On the grid draw the graph of $y = \dfrac{12}{x}$ for values of x from 1 to 12.

(You do not need to work out further points.)

(3 marks)

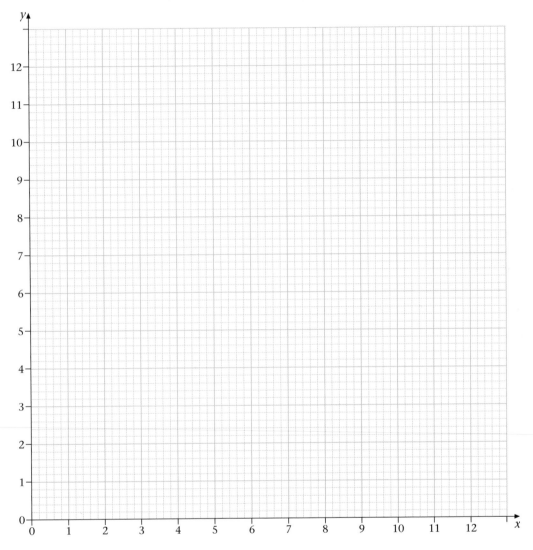

c Using your graph, or otherwise, solve the equation $\dfrac{12}{x} = 7$.

....

....

Answer .... *(1 mark)*

317

15 Using ruler and compasses only, construct the angle bisector of angle *A*.
   You **must** show clearly all your construction arcs.

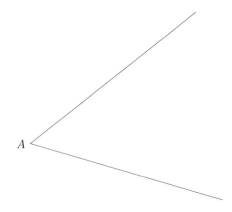

(3 marks)

16 A rectangle has length 4.5 cm and width 1 cm.
   Work out the area of the rectangle,
   giving your answer in mm².

   ....

   ....

   Answer... *(2 marks)*

17 $y = 4x - 3$ is the equation of a straight line.

   a Write down the *y*-intercept of the line.

   Answer .... *(1 mark)*

   b Write down the gradient of the line.

   Answer .... *(1 mark)*

18 Make *x* the subject of $y = \sqrt{x + 4}$

   ....

   ....

   Answer... *(2 marks)*

19 a In the diagram, *O* is the centre of the circle.

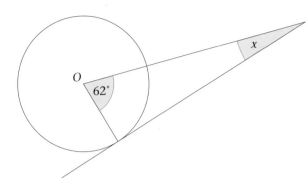

Not drawn accurately

Calculate the value of *x*.

....

....

Answer... *(2 marks)*

b

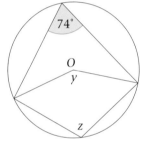

Not drawn accurately

i Work out the value of *y*.

....

....

Answer... *(1 mark)*

ii Work out the value of *z*.

....

....

Answer... *(1 mark)*

20 Solve the simultaneous equations:

$2x + y = 7$

$3x - 2y = 7$

You **must** show your working.
Do **not** use trial and improvement.

....

....

....

....

....

....

Answer... (4 marks)

Paper 2   Calculator   1 hour 15 minutes

Total marks of 70 for Paper 2.

1 Work out $5^3 + 4^2$.

....

....

Answer... (2 marks)

2 Sasha draws a rough sketch of a triangle with sides $AB = 80$ cm,
$BC = 150$ cm and angle $ABC = 60°$.

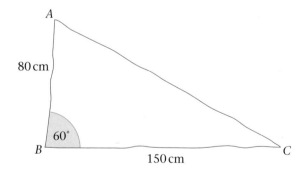

Using ruler and compasses only, make an accurate scale drawing of the triangle.
Use a scale of 1 cm to represent 20 cm.
You **must** show your construction lines clearly. *(3 marks)*

3 Solve the equations

   a $5x - 8 = 12$

....

....

           Answer.. *(2 marks)*

   b $8(y - 3) = 32$

....

....

           Answer.. *(3 marks)*

4 A square has sides of length 4.8 cm.

Not drawn accurately

4.8 cm

4.8 cm

   a Calculate the area of the square.

....

....

           Answer.. *(2 marks)*

   b A rectangle has the same area as the square.
      The length of the rectangle is 9.6 cm.

Not drawn accurately

9.6cm

   Explain why the height of the rectangle must be 2.4 cm.

....

.... *(1 mark)*

5 This shape is made from a square with side length 3.5 cm and a rectangle of length 8.4 cm and height 2.2 cm.

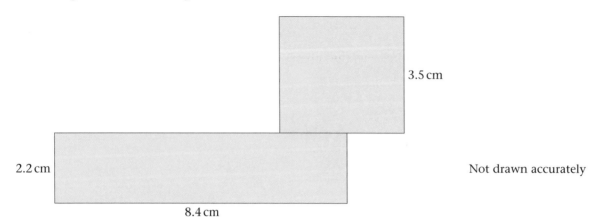

3.5 cm

2.2 cm

Not drawn accurately

8.4 cm

Calculate the total area of this shape.

....

....

Answer.. *(2 marks)*

6 a Complete the table of values for the graph $x - y = 5$.

x	0	1	2	3	4	5
y	−5			−2		

(2 marks)

b On the grid, draw the graph of $x - y = 5$.

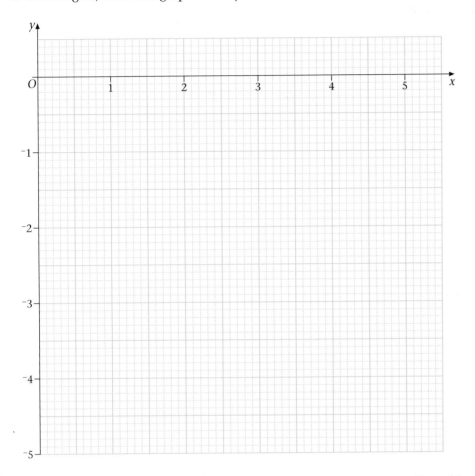

(1 mark)

c Q is a point with coordinates $(1, -3)$.

Is the point Q above the line, on the line, or below the line?

Answer .... *(1 mark)*

7 Triangle PQR is isosceles. $PQ = PR$

Calculate the size of the angle marked x.

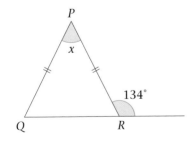

Not drawn
accurately

....

....

Answer .... *(3 marks)*

8 The diagram shows a triangular prism.

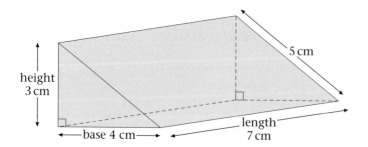

On cm square paper, draw an accurate net for this prism.

(3 marks)

9 Part of a number grid is shown below.

1	2	3	4	5	6	7	8	9	10
11	12	13	14	15	16	17	18	19	20
21	22	23	24	25	26	27	28	29	30
31	32	33	34	35	36	37	38	39	40
41	42	43	44	45	46	47	48	49	50

The grid selected in green is called S_{13} because it is a 2×2 square with the number 13 in the top left hand corner of the square.
The sum of the numbers in this square is 74.

a This is S_n.

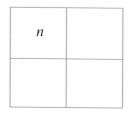

Fill in the empty boxes of S_n. *(2 marks)*

b Find the sum of all of the numbers in S_n in terms of n.
Give your answer in its simplest form.

....

....

Answer... *(2 marks)*

c Explain why the sum of the numbers in S_n can never be odd.

....

.... *(2 marks)*

10 Amy has a tricycle.

The radius of the front wheel is 10 cm.

The radius of one of the rear wheels is 6 cm.

Calculate how much greater the circumference of the front wheel is than one of the rear wheels.

....

....

....

....

Answer.. *(4 marks)*

11 Work out the value of angle x.

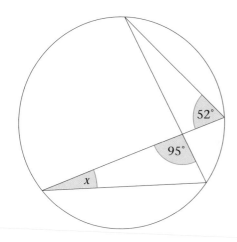

....

....

....

Answer.. *(3 marks)*

12 Jono used trial and improvement to find a solution to this equation:
$x^3 - 2x = 10$

x	$x^3 - 2x$	Comment
2	4	too small
3	21	too big

Continue the table to find a solution to the equation.
Give your answer to 1 decimal place.

Answer... *(3 marks)*

13 Wayne says 'this triangle is right angled.'
Is he correct? You **must** show your working.

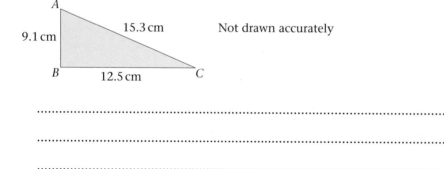

Not drawn accurately

....

....

....

.... *(3 marks)*

14 A circular garden pond has a diameter of 4 m.
The leaves of a water lily cover the surface of the water in a circle that has
a diameter of 1 m.

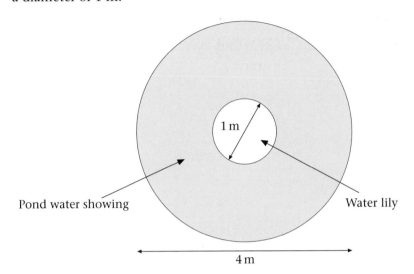

Pond water showing

1 m

Water lily

4 m

Calculate the surface area of the pond water that is showing.

....

....

....

....

....

....

Answer... *(5 marks)*

15 Solve the equation
$3(x - 6) = x - 4$

....

....

....

Answer... *(3 marks)*

16 In the expressions below, the letters w, x, y and z each represent lengths. State whether each expression represents length, area, volume or none of these.

a $w + x + y + z$

Answer ....

b $y^2 z$

Answer ....

c x^4

Answer ....

(2 marks)

17 a List the integer values of p such that $-6 < 2p \leqslant 8$.

....

Answer... *(3 marks)*

b Find the equation of the line RS.

....

....

Answer... *(2 marks)*

c On the grid shade the region that satisfies these inequalities.

i $y \leqslant x + 1$

ii $x < 4$

iii $y > 1$

(2 marks)

18 In the diagram, $BC = 5$ m.
Angle $BCD = 40°$
Angle $BAD = 35°$
BD is perpendicular to AC.

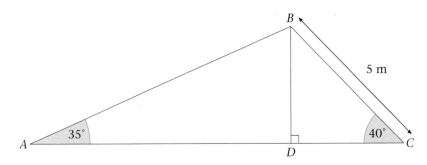

Calculate the length of AD.
You **must** show your working.

....

....

....

....

....

....

Answer... *(5 marks)*

19 a Factorise
$x^2 - 5x - 36$

....

....

....

Answer... *(2 marks)*

b Hence, or otherwise, solve the equation
$x^2 - 5x - 36 = 0$

....

....

....

Answer .... *(1 mark)*

Answers

Chapter 1 Revision of number skills

Practice questions

1 16, 49, 121
2 a 49   **b** 121   **c** 196
3 a 88.36   **b** 171.61   **c** 0.09
4 77
5 29
6 +12 and −12
7 64, 125, 1000
8 a 343   **b** 512   **c** 27 000
9 a 68.921   **b** 1092.727   **c** 0.125
10 $\sqrt[3]{314\,432}$, 68.12, 68.5, $68\frac{2}{3}$, 4.1^3
11 121
12 612
13 a 80.3   **b** 73.4   **c** 31.1   **d** 43.8
14 a 100   **b** 45   **c** 56   **d** 230
15 268 cm^3 or 270 cm^3
16 a i 429.3   **ii** 4.293×10^2
      b i 0.084   **ii** 8.4×10^{-2}
      c i 600 000   **ii** 6×10^5
17 £105
18 756 kg
19 104 g or 100 g
20 57.8 litres or 58 litres

Chapter 2 Equivalent fractions

Practice question 1

1 a $\frac{3}{5}$ **b** $\frac{4}{5}$ **c** $\frac{2}{5}$ **d** $\frac{3}{4}$
   e $\frac{4}{7}$ **f** $\frac{2}{3}$ **g** $\frac{1}{4}$ **h** $\frac{2}{3}$

Practice questions 2

1 a $\frac{1}{4}, \frac{1}{3}, \frac{2}{5}$ **b** $\frac{2}{3}, \frac{7}{10}, \frac{3}{4}$ **c** $\frac{4}{9}, \frac{1}{2}, \frac{5}{6}$ **d** $\frac{3}{5}, \frac{19}{30}, \frac{13}{20}, \frac{7}{10}$
2 $\frac{3}{10}$ (0.3) is 0.2 away from 0.5, $\frac{3}{4}$ (0.75) is 0.25 away from 0.5. Therefore $\frac{3}{10}$ is the closest.
3 $\frac{14}{15}$ is furthest away from 1 as $\frac{1}{15}$ is greater than both $\frac{1}{20}$ and $\frac{1}{25}$.

Practice exam question

1 $\frac{19}{40}$ because $\frac{4}{10} = \frac{48}{120}$, $\frac{9}{20} = \frac{54}{120}$, $\frac{14}{30} = \frac{56}{120}$, $\frac{19}{40} = \frac{57}{120}$, $\frac{1}{2} = \frac{60}{120}$

Chapter 3 Collecting like terms

Practice question

1 a $3a - 2b$ **b** $10c + d$ **c** $-e - 3f$ **d** $-g$ **e** $-h$
   f $-k + m$ **g** $p - q$ **h** $-9r$ **i** $5xy + 2x$ **j** $6w^2 + w$

Practice exam questions

1 $p - q$
2 $5x + 10y$
3 $15p - 3q$
4 $7a + 2b + 4ab$

Chapter 4 Rules of indices

Practice question

1 a a^{10} **b** b^{15} **c** c^5 **d** d^2 **e** 1
   f $\frac{1}{f^2}$ or f^{-2} **g** g^6 **h** h^{16} **i** i^3 **j** $8j^4k^5$
   k $21l^7m^5p^2$ **l** $2q$ **m** $4t^4$ **n** $\frac{16}{3u^2}$ **o** $6w^2z^4$

Practice exam questions

1 a a^4 **b** a^6
2 a t^9 **b** t^3 **c** t^4
3 $3a^6b^2$
4 a i a^8 **ii** a^2 **iii** a^{15} **b i** $(a^5)^3$ **ii** $a^5 \div a^3$

Chapter 5 Expanding brackets

Practice question 1

1 a $6x^2 + 12xy - 3xz$ **b** $2t^2 + 3rt$ **c** $4pq - 8p^2$
   d $-6x^2 - 2xy + 2xz$ **e** $x^3 + 3x^2 - 4x$

Practice questions 2

1 a $7x - 15$ **b** $6x - 10$ **c** $13x - 35$ **d** $11p - 8p^2 - 7$
2 a $10g^2 + 9g$ **b** $6y^2 - 24y - 3$ **c** $-2x - 2$ **d** $19a - 19$

Practice question 3

1 a $18a^4 - 12ab$ **b** $14x^5 - 7x^3y$ **c** $20y^2 - 12y^3$
   d $10x^3 + 15x^2y$ **e** $12t^4 - 8t^3 + 12t^2 - 24t$

Practice question 4

1 a $x^2 + 13x + 42$ **b** $x^2 + 6x + 8$ **c** $x^2 + 2x - 24$
   d $x^2 - 6x + 5$ **e** $x^2 + 8x - 9$

Practice question 5

1 a $t^2 + 10t + 16$ **b** $u^2 - 7u + 12$ **c** $6x^2 + x - 2$

Practice questions 6

1 a $x^2 - x - 12$ **b** $x^2 - x - 20$
2 a $2x^2 - 2x - 4$ **b** $8x^2 + 22x + 15$ **c** $15p^2 - 41p + 28$
3 a $x^2 + 10x + 25$ **b** $x^2 + 18x + 81$
   c $x^2 - 6x + 9$ **d** $x^2 - 12x + 36$
4 a $4x^2 + 4x + 1$ **b** $9x^2 + 12x + 4$ **c** $4x^2 - 16x + 16$
   d $4x^2 - 12x + 9$
5 a $x^2 + 2xy + y^2$ **b** $4x^2 - 4xy + y^2$
   c $9x^2 + 12xy + 4y^2$ **d** $9x^2 - 24xy + 16y^2$

Practice exam questions

1 $6c - 15$
2 a $10x + 18$ **b** $2x + 3y$
3 a $6x + 4xy$ **b** $4a^5 - 4a^3b$
4 a $20 - 3x$ **b** $12x^2 + 5xy - 2y^2$
5 $4x^2 + 12xy + 9y^2$

Chapter 6 Changing the subject of a formula

Practice questions

1 $k = \dfrac{2n + m}{8}$ 2 $x = \dfrac{y + 10}{3}$

3 a $s = \dfrac{v^2 - u^2}{2a}$ b $a = \dfrac{v^2 - u^2}{2s}$ c $u = \pm\sqrt{v^2 - 2as}$

4 $y = \tfrac{3}{2}x - 2$ 5 $b = \pm\sqrt{15 - c}$ 6 $m = \dfrac{5n + 6}{3}$

Practice exam questions

1 $x = \dfrac{y + 6}{5}$ 2 $y = \dfrac{40 - x}{8}$ 3 $x = \dfrac{y - c}{m}$

4 $x = \pm\sqrt{w - y}$ 5 $q = \pm\sqrt{p - r}$ 6 $t = \dfrac{4W - 3}{5}$

7 $x = \pm\sqrt{16 - k}$ 8 $p = \dfrac{7r - 12}{4}$

Chapter 7 Substitution into a formula

Practice questions 1

1 a $\tfrac{1}{2}$ b 3   c 1   d -2
2 a $p = 20$ b $l = 11$
3 a $P = 1.1125$ b $R = 4.4$ (to 2 s.f.)

Practice question 2

1 a $f(1) = 1$, $f(-1) = 1$, $f(3) = 9$
   b $f(1) = -2$, $f(-1) = -6$, $f(3) = 2$
   c $f(1) = 7$, $f(-1) = 7$, $f(3) = 23$
   d $f(1) = 1$, $f(-1) = -1$, $f(3) = \tfrac{1}{3}$

Practice exam questions

1 6   2 a 111.8   b 140   3 9.2   4 3   5 71.8   6 2.2
7 1   8 a i £32.16   ii 150 miles   b £153.48
9 9.26 (to 3 significant figures)   10 0   11 3.444

Chapter 8 Number patterns and sequences

Practice questions 1

1 a $4, -1$ Subtract 5
   b $-8, -9$ Subtract 5, then 4, then 3, etc.
   c $-1, -3\tfrac{1}{2}$ Subtract $2\tfrac{1}{2}$
   d $-19, -32$ Subtract 5, then 7, then 9, etc.
2 13 sticks. Increases by 2 each time, or double pattern number and add 1.
3 $(1 + 2 + 3 + 4)^2 = 1^3 + 2^3 + 3^3 + 4^3$
   $(1 + 2 + 3 + 4 + 5)^2 = 1^3 + 2^3 + 3^3 + 4^3 + 5^3$
4 a $9^2 - 5^2 = 4 \times 14$
   b $21^2 - 11^2 = 10 \times 32$
5 a $1 + 2 + 3 + 4 + 5 = \tfrac{5 \times 6}{2}$
   b $1 + 2 + 3 + 4 + 5 + 6 + 7 + 8 + 9 + 10 + 11 = \tfrac{11 \times 12}{2}$

Practice questions 2

1 a $5n - 2$ b $4n - 2$ c $2n + 8$ d $4n + 1$
2 a $9^2 - 5 \times 9 + 6 = 7 \times 6$
   b $(n + 3)^2 - 5 \times (n + 3) + 6 = (n + 1) \times n$

Practice exam questions

1 a 31   b -17
   b 23   c $3n + 2$ d 20th pattern

2 a

Pattern	1	2	3	4	5
Number of dots	5	8	11	14	17

3 a 19   b $x + 4$ c $x - 4$
4 $-2, -8$
5 a i 5   ii 18
   b i $n + 1$ ii $4n + 2$
6 a $4^2 + 8 = 6^2 - 12$
   b i $4^2 + 8 = 4 \times 6$ ii $n^2 + 2n = n \times (n + 2)$
7 $10 \times 4 - 5 = 7 \times 5$
8 a $10^2 - 9^2 = 19$ b $r^2 - (r - 1)^2 = 2r - 1$

Chapter 9 Angle facts 1

Practice question 1

1 a $a = 146°$, $b = 34°$ b $c = 19°$ c $d = 159°$
   d $e = 153°$ e $f = 55°$, $g = 77°$, $h = 48°$
   f $i = 60°$

Practice questions 2

1 a i 72°   ii 108°   iii 540°
   b i 60°   ii 120°   iii 720°
   c i 45°   ii 135°   iii 1080°
   d i 36°   ii 144°   iii 1440°

2

Number of sides	Exterior angle	Interior angle	Sum of interior angles
7	51.4°	128.6°	900.2°
9	40°	140°	1260°
11	32.7°	147.3°	1620.3°

Practice question 3

1 a

   b

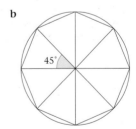

   c

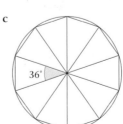

Practice exam questions

1   $x = 118°$
2   $x = 63°$
3   **a** $x = 50°$, $y = 40°$
    b i Kite   **ii** $p = 100°$, $q = 118°$
4   $z = 80°$
5   140° **6 a** 45° **b** 67.5°
7   105°

Chapter 10 Constructing triangles

Practice question

Accurate drawings of triangles

Chapter 11 Circles

Practice questions 1

1   **a** 8π cm   **b** 34π cm
    c 2π m   **d** 4.8π cm
2   **a** 16 cm   **b** 4.5 cm   **c** 3.0 cm   **d** 72 cm
3   **a** 14 cm   **b** 38 cm   **c** 18 cm   **d** 110 cm

Practice questions 2

1   **a i** 25π cm^2 **ii** 79 cm^2 **b i** 529π cm^2 **ii** 1700 cm^2
    c i 16π m^2 **ii** 50 m^2 **d i** 73.96π cm^2 **ii** 230 cm^2
2   **a** 6.2 cm   **b** 4.4 cm   **c** 5.1 cm   **d** 38 cm

Practice exam questions

1   **a** 81.7 cm   **b** 19.6 cm^2
2   **a** 112   **b** 31.4 cm

Chapter 12 Plans and elevations

Practice questions

1   **a** iv   **b** vi   **c** ii   **d** v   **e** iii

2   **a**   Plan view

   Front elevation   Side elevation

b   Plan view

   Front elevation   Side elevation

c   Plan view

   Front elevation   Side elevation

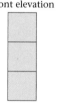

d   Plan view

   Front elevation   Side elevation

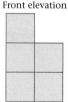   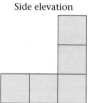

e   Plan view

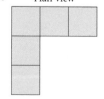

   Front elevation   Side elevation

f

Plan view

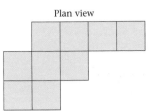

Front elevation

Side elevation

3

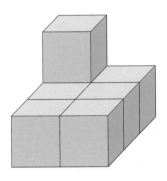

Chapter 13 Solving linear equations

Practice question

1 **a** $r = 3$ **b** $s = 0.5$ **c** $t = 48$ **d** $x = 3$
    e $y = 6$ **f** $z = 14$ **g** $a = -6.5$ **h** $b = 2$
    i $m = -3$ **j** $x = 16$ **k** $h = -3$

Practice exam questions

1 **a** $x = 1.5$ **b** $x = 0.5$ **2** $x = 0.25$
3 $x = 5$ **4 a** $x = \frac{1}{3}$ **b** $x = 22$ **5** $x = 6$
6 **a** $x = 3$ **b** $x = 2$ **7 a** $x = -1$ **b** $x = 2.8$

Chapter 14 Forming and solving linear equations

Practice questions 1

1 **a** 13   **b** 17   **c** 52   **d** 92
    e 31   **f** 27   **g** 58
2 32

Practice questions 2

1 14 cm
2 **a** $30 + x + (x + 20) + 3x + (x + 70)$ or $6x + 120$
    b $30 + x + (x + 20) + 3x + (x + 70) = 540$ or $6x + 120 = 540$
    c 70°
    d 210°

Practice exam questions

1 **a** 6 **b** 12 **2 a** 8 **b** 15 **c** $12x - 2$ **d** ×3, +2
3 23° F **4 a** $5x + 1$ **b i** $x = 3$ **ii** 5 **5** $x = 6$
6 **a** $16x + 4(2x + 3) = 24x + 12$ **b** $24x + 12 = 132$, $x = 5$

Chapter 15 Linear graphs

Practice questions 1

1

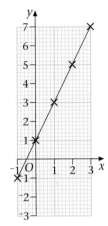

2

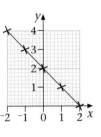

3

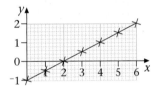

Practice questions 2

1 a

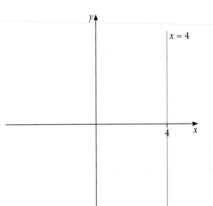

b

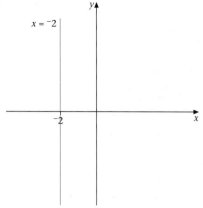

c

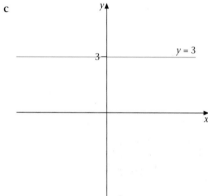

d

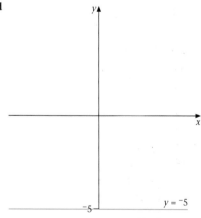

2 **a** Gradient $m = 2$, y-intercept $c = 1$
   b Gradient $m = 5$, y-intercept $c = 0$
   c Gradient $m = \frac{1}{4}$, y-intercept $c = \frac{5}{4}$
   d Gradient $m = -\frac{3}{7}$, y-intercept $c = \frac{4}{7}$
   e Gradient $m = 2$, y-intercept $c = -3$
3 **a** $y = 4x + 1$ **b** $y = 3x$ **c** $y = -\frac{5}{4}x + 5$ **d** $y = -\frac{2}{3}x + 4$

Practice questions 3

1 **a** $y = 2x + 1$ and $y = 2x + \frac{7}{2}$, parallel
   b $y = 4x + \frac{1}{2}$ and $y = \frac{1}{4}x - 2$, not parallel
   c $y = \frac{1}{2}x + \frac{2}{3}$ and $y = \frac{1}{2}x + 3$, parallel
   d $y = \frac{1}{2}x + 1$ and $y = \frac{1}{2}x + 2$, parallel

2 **a** $m = 2$, $c = 1$; $m = 2$, $c = \frac{7}{2}$

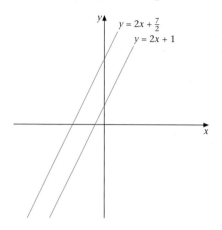

b $m = 4$, $c = \frac{1}{2}$; $m = \frac{1}{4}$, $c = -2$

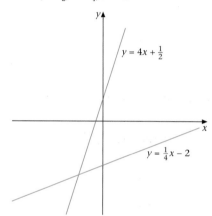

c $m = \frac{1}{2}$, $c = \frac{2}{3}$; $m = \frac{1}{2}$, $c = 3$

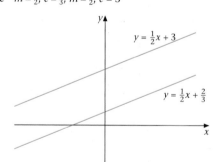

d $m = \frac{1}{2}$, $c = 1$; $m = \frac{1}{2}$, $c = 2$

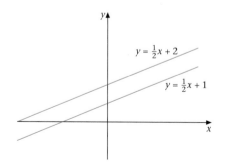

Practice exam questions

1 $\frac{4}{5}$

2 a

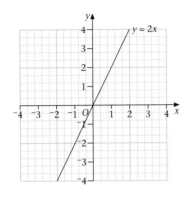

b (−1, −2)

3 a 12 cm **b** 2 **c** 10 kg

Chapter 16 Transformations

Practice questions 1

1 a Translation $\begin{pmatrix} 2 \\ 3 \end{pmatrix}$ **b** Translation $\begin{pmatrix} -7 \\ 1 \end{pmatrix}$

c Translation $\begin{pmatrix} -5 \\ -4 \end{pmatrix}$ **d** Translation $\begin{pmatrix} 0 \\ -5 \end{pmatrix}$

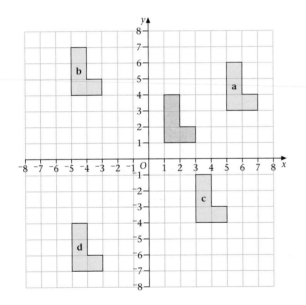

Practice questions 2

1

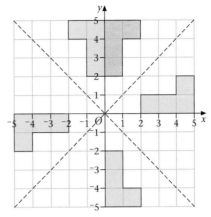

2 a Reflection in the line $x = 4$
b Reflection in the line $y = x$

Practice questions 3

1

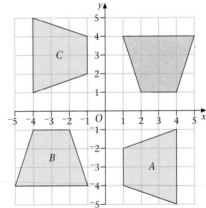

2

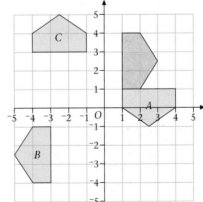

3 a Rotation 90° clockwise about (1, 3)
b Rotation 90° anticlockwise about (1, 3)

Practice questions 4

1

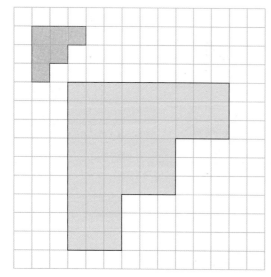

2 Enlargement, scale factor 2, centre of enlargement (1, 10)

3

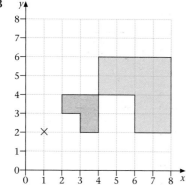

4 10.35 m

Practice question 5

1 **a** 3 **b** 5 **c** 4 **d** 6 **e** 1

Practice exam questions

1 **a** (0, 1) **b** $\begin{pmatrix} 3 \\ -6 \end{pmatrix}$

2 **a** Reflection in the *y*-axis
b Rotation 180° about (3, 0)

c

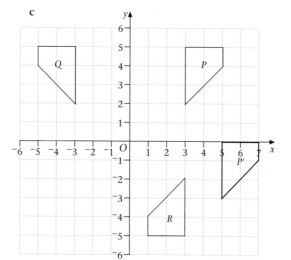

3

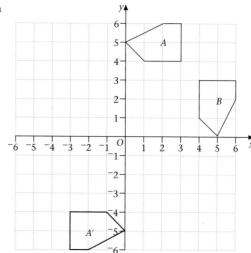

4 **a** Reflection in the line *x* = 3
b Rotation 180° about (3, 0)

5 **a**

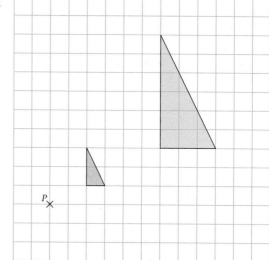

b Reflection in the line *y* = *x*

6 a

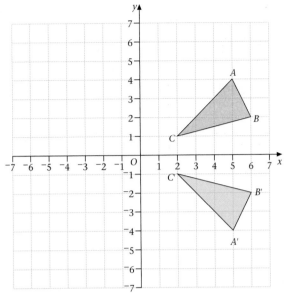

b i (−4, −6) **ii** (25, −14)

7 (−4, −1)

8 Rotation 180° about (3, 0)

9 a Rotation 90° anticlockwise about (0, 0)

b Reflection in the line $y = x$

c

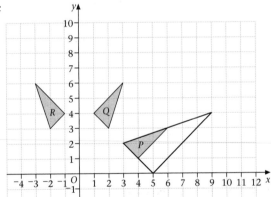

10

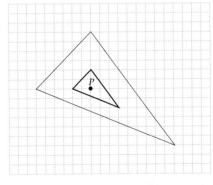

11

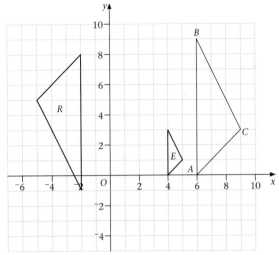

12 a (3, 6) **b** (3, 2)

13 a

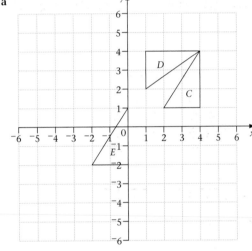

b Reflection in the line $y = x$

14 Rotation 180° about (0, 0)

Chapter 17 Angle facts 2

Practice question

1 a $a = 136°$ (alternate angles), $b = 44°$ (angles on a straight line add up to 180°)

b $c = 52°$ (corresponding angles), $d = 52°$ (vertically opposite angles), $e = 128°$ (angles on a straight line add up to 180°)

c $f = 110°$ (vertically opposite angles),
$g = 129°$ (angles on a straight line add up to 180°),
$h = 51°$ (alternate angles),
$i = 59°$ (angles in a triangle add up to 180°)

d $j = 90°$ (allied angles add up to 180°), $k = 75°$ (vertically opposite angles)

Practice exam questions

1 $x = 63°$, $y = 27°$
2 $p = 25°$, $q = 105°$
3 **a** $f = 80°$
   b $g = 140°$
   c

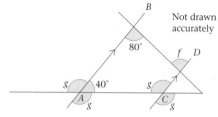

Not drawn accurately

Practice exam questions

1 **a** $2a(3 + 5b - 4ab)$ **b** $(a - 3)(a + 3)$ **2** $3x(x - 2)$
3 $5xy(x + 3y^2)$ **4** $(x - 7)(x - 3)$ **5** $3a(1 - 9a)$
6 **a** $3(pq - 2r)$ **b** $(c - 4)(c - 5)$ **7** $2x(x + 2)$

Chapter 19 Cancelling common factors

Practice questions

1 **a** a **b** y^2 **c** $\dfrac{b}{a}$ **d** $\dfrac{x}{(x - 2)}$

   e $\dfrac{y(y + 1)}{2}$ **f** $3(x - 1)$ **g** $\dfrac{3y^2}{x}$

Practice exam question

1 $\dfrac{a^4}{c}$

Chapter 18 Factorising

Practice question 1

1 **a** $3(x + 4y)$ **b** $4(s - 6t)$ **c** $3(5a + b)$
   d $6(x - 3y)$ **e** $7(2p + 5q)$ **f** $2(5e - 4s)$

Practice question 2

1 **a** $p(3 + 4q)$ **b** $r(5 - 8q)$ **c** $s(6 + 7t)$
   d $x(x - 16)$ **e** $x(3x + 11)$ **f** $x(5x + 12)$

Practice question 3

1 **a** $5p(1 + 4q)$ **b** $2x(x + 3)$ **c** $7t(s - 2)$
   d $3a(4a + b)$ **e** $4r(2 - q)$ **f** $3x(3x + 2)$

Practice question 4

1 **a** $xy(5 - 12y)$ **b** $4pq(1 + 5p)$ **c** $3rs(s^2 - 2r)$
   d $cd(d - c)$ **e** $5xyz(1 + 3xz^2)$

Practice questions 5

1 **a** $(x + 2)(x + 1)$ **b** $(x + 3)(x + 1)$ **c** $(x + 3)(x + 5)$
   d $(x + 6)(x + 1)$ **e** $(x - 4)(x - 6)$ **f** $(x - 2)(x - 3)$
2 **a** $(x + 3)(x - 1)$ **b** $(x + 6)(x - 1)$ **c** $(x + 5)(x - 2)$
   d $(x + 9)(x - 2)$ **e** $(x + 8)(x - 3)$
3 **a** $(x - 3)(x + 1)$ **b** $(x + 2)(x - 7)$ **c** $(x + 3)(x - 6)$
   d $(x + 2)(x - 6)$ **e** $(x + 3)(x - 10)$
4 **a** $(y + 3)(y - 5)$ **b** $(p + 7)(p - 4)$ **c** $(t + 3)(t + 6)$
   d $(z + 3)(z - 7)$

Practice question 6

1 **a** $(x + 2)^2$ **b** $(x + 3)^2$ **c** $(x + 5)^2$
   d $(x - 2)^2$ **e** $(x - 4)^2$ **f** $(x - 6)^2$

Practice question 7

1 **a** $(x + 1)(x - 1)$ **b** $(y + 4)(y - 4)$ **c** $(z + 15)(z - 15)$
   d $(t + 8)(t - 8)$ **e** $(r + 9)(r - 9)$ **f** $(s + 11)(s - 11)$
   g $(6 - x)(6 + x)$ **h** $(5 - p)(5 + p)$ **i** $(1 + q)(1 - q)$

Practice questions 8

1 **a** 200   **b** 8000   **c** 640
2 **a** 9600   **b** 7800   **c** 2200

Chapter 20 Solving quadratic equations

Practice questions 1

1 **a** $x = -2$ and $x = -5$ **b** $x = -4$ and $x = 2$
   c $x = -6$ and $x = 1$ **d** $x = -2$ and $x = -3$
   e $x = 9$ and $x = -3$ **f** $x = 10$ and $x = -1$
2 **a** $y = -10$ and $y = 1$ **b** $p = 9$ and $p = -1$
   c $t = -4$ and $t = 1$ **d** $s = 3$ and $s = -1$
   e $z = 2$ and $z = 6$

Practice questions 2

1 **a** $x = 0$ and $x = 8$ **b** $x = 0$ and $x = -2$
   c $x = 0$ and $x = 5$ **d** $x = 0$ and $x = -3$
   e $x = 0$ and $x = 1$
2 **a** $y = 0$ and $y = 12$ **b** $p = 0$ and $p = 7$
   c $s = 0$ and $s = -6$ **d** $t = 0$ and $t = -23$
   e $z = 0$ and $z = 15$

Practice question 3

1 **a** $x = \pm 9$ **b** $x = \pm 4$ **c** $x = \pm 12$
   d $x = \pm 13$ **e** $x = \pm 15$

Practice questions 4

1 **a** $(x + 6)(x + 2) = 77$
     $x = +5$ or -13
     $x = 13$ would make the lengths negative so $x = +5$.
   b 11 cm and 7 cm
2 **a** $x^2 + 6x - 40 = 0$
     $x = +4$ or -10
     $x = +4$ as $x = -10$ would give a negative length.
   b 7 cm
3 **a** $x^2 - 2x - 8 = 0$
     $x = +4$ or -2
     $x = +4$ as it cannot be negative.
   b 4 cm, 10 cm
   c P 28 cm, Q 26 cm so Q's perimeter is 2 cm longer than P's.

Practice exam questions

1 a The area is $4x^2 + 4x = 48$.

To simplify divide each side by 4:

$x^2 + x = 12$

Subtract 12 from each side:

$x^2 + x - 12 = 0$

b $x = -4$ or 3, x cannot be negative so $x = +3$.

2 a 0   **b** $(x - 4)(x + 3)$ **c** $x = 4$ or -3

3 $x = -2$ or -6

4 a $(x - 7)(x + 4)$ **b** $x = 7$ or -4

Chapter 21 Quadratic graphs

Practice question

1 a i $y = x^2 - 6$

x	−4	−3	−2	−1	0	1	2	3	4
y	10	3	−2	−5	−6	−5	−2	3	10

ii

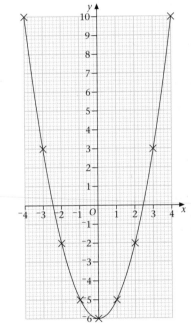

iii The line of symmetry is $x = 0$ (the y-axis).

iv When $y = 0$

$x^2 - 6 = 0$

$x = \pm 2.45$ (± 2.4 or ± 2.5 would be acceptable answers from a graph)

b i $y = x^2 - 2x - 3$

x	−4	−3	−2	−1	0	1	2	3	4
y	21	12	5	0	−3	−4	−3	0	5

ii

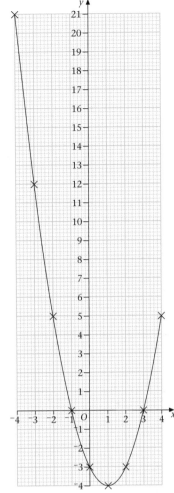

iii The line of symmetry is $x = 1$ if the x-axis is extended to +6.

iv When $y = 0$

$x^2 - 2x - 3 = 0$

$x = -1$ or $+3$

c i $y = 8 - x^2$

x	−4	−3	−2	−1	0	1	2	3	4
y	−8	−1	4	7	8	7	4	−1	−8

ii

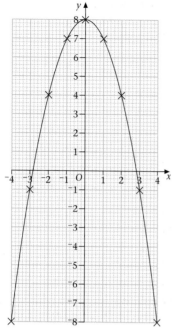

iii The line of symmetry is $x = 0$ (the y-axis).

iv When $y = 0$
$8 - x^2 = 0$
$x = \pm 2.8$

d i $y = 7 - \frac{1}{2}x^2$

x	−4	−3	−2	−1	0	1	2	3	4
y	−1	2.5	5	6.5	7	6.5	5	2.5	−1

ii

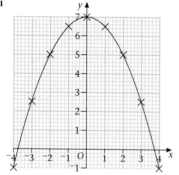

iii The line of symmetry is $x = 0$ (the y-axis).

iv When $y = 0$
$7 - \frac{1}{2}x^2 = 0$
$x = \pm 3.7$

Practice exam questions

1 a $y = x^2 + 3$

x	−2	−1	0	1	2	3
y	7	4	3	4	7	12

b

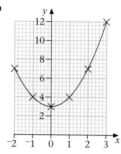

2 a i $y = x^2 - 7$

x	−3	−2	−1	0	1	2	3
y	2	−3	−6	−7	−6	−3	2

ii

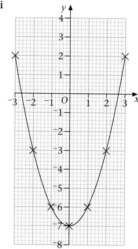

b $x = \pm 2.6$

c The minimum value of y is −7.

3 a $y = 5 + x - x^2$

x	−3	−2	−1	0	0.5	1	2	3
y	−7	−1	3	5	5.25	5	3	−1

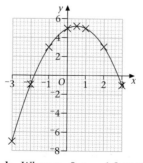

b When $y = 0$, $x = -1.8$ or $+2.8$.

4 a $y = x^2 - 2x - 5$

x	–3	–2	–1	0	1	2	3	4
y	10	3	–2	–5	–6	–5	–2	3

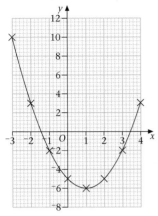

b $x^2 - 2x - 5 = 0$ when $x = -1.4$ or $+ 3.4$.

5 a and **b**

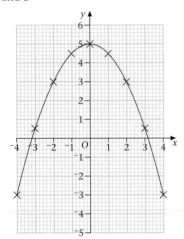

c $x = \pm 3.2$

Chapter 22 Interpreting graphs

Practice questions 1

1 a

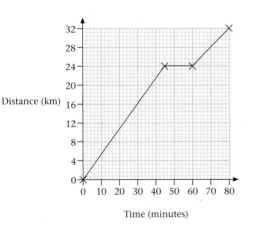

b Average speed $= \dfrac{\text{total distance}}{\text{total time}}$

(total time is 80 minutes or $1\frac{1}{3}$ hours)

$= \dfrac{32}{1\frac{1}{3}}$

$= 24$ km/h

2 a 42 minutes

b 8 mph

c 6.09 pm

Practice questions 2

1

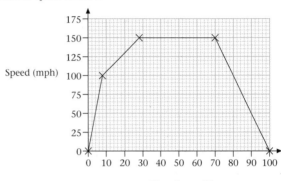

2 The runner sets off from rest and gradually increases his speed to 6 ms⁻¹ after 2 minutes.
He then gradually increases his speed to 8 ms⁻¹ over the next 8 minutes.
He maintains this speed for the next 20 minutes.
From 30 to 50 minutes the runner gradually increases his speed from 8 ms⁻¹ to 9.2 ms⁻¹.
He continues to run at 9.2 ms⁻¹ for 4 minutes and then slows gradually to 9 ms⁻¹ for the last 6 minutes of the first hour.

Practice question 3

1 a i 0 °C   **ii** 10 °C   **iii** 38 °C   **iv** 100 °C
  b i 68 °F   **ii** 86 °F   **iii** 41 °F   **iv** 14 °F

Practice questions 4

1

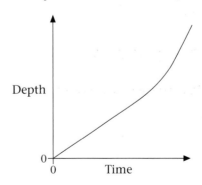

2 B, the depth will drop at a constant rate.

Answers

Practice question 5

1

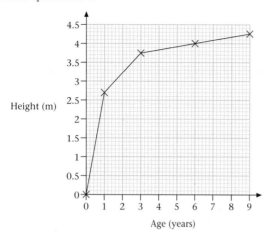

Height (m) vs Age (years)

Practice exam questions

1 **a** 1045 **b** 2.2 km **c** 3.6 km/h
2 **a** 30 minutes **b** 16 mph
  c *CD* because it is the steepest part of the graphs.
  d 4 pm
3 **a** 2.8 km
  b Jack and Jill stayed at Paildon from 1108 till 1140.
  c They started to walk slower than the first part of their walk.
  d 4.2 km/h
4 **a** 5.4 km **b** 9.36
  c

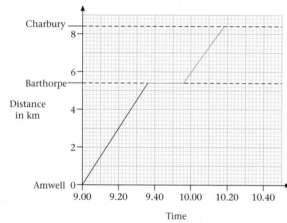

5 **a** B **b** D
  c

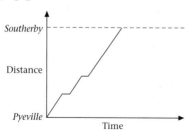

6 **a** Nina **b** They were neck and neck (level).
  c Nina stopped for 20 s. **d** Nina

7 **a**

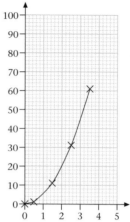

  b 3.1 s
  c Extend the curve and read at 4 s.

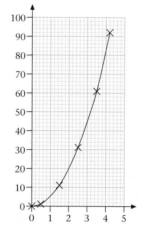

  About 83 m
8 32 km/h = 20 mph
  so 96 km/h = 60 mph

Chapter 23 Pythagoras' theorem

Practice question 1

1 **a** $x = 10$ cm **b** $b = 12$ cm **c** $a = 9$ cm **d** $b = 12$ cm
2 **a** 150 cm **b** 4 cm **c** 50 cm **d** 26.2 cm

Practice question 2

1 **a** Yes (3, 4, 5), *B*
  b No
  c No
  d Yes, because $9.5^2 = 7.3^2 + 6.08^2$, *X*
  e Yes, because $12.5^2 = 3.5^2 + 12^2$, *M*

Practice questions 3

1 **a** 5 units **b** 3.61 units **c** 13 units
  d 6 units **e** 7.07 units
2 11.2 units

Practice exam questions

1 24.2 m
2 **a** A(0, 3)
   b B(2, 0) so AB = $\sqrt{13}$ = 3.61 units
3 5.61 m
4 **a** x = 19.2 km
   b y = 9.80 km
5 $9^2 + 40^2 = 1681$
   $\sqrt{1681} = 41$
   Triangle *ABC* is a right-angled triangle.

Practice question 1

1

	opposite	adjacent	hypotenuse
a	*BC*	*AC*	*AB*
b	*BC*	*AC*	*AB*
c	*BC*	*AB*	*AC*
d	*BC*	*AB*	*AC*
e	*BC*	*AB*	*AC*

Practice questions 2

1 z = 6.10 cm   2 a = 13.2 cm   3 p = 15.5 cm
4 t = 3.36 m   5 b = 18.6 cm   6 g = 14.9 cm
7 j = 23.5 cm   8 k = 1.24 km   9 r = 5.03 km
10 s = 1.26 cm   11 t = 3.79 m   12 s = 34.5 m
13 ladder = 3.77 m 14 rope = 4.14 m 15 ladder = 4.73 m
16 rope = 7.25 m   17 x = 6.25 m

Practice questions 3

1 **a** 31.4°   **b** 47.8°   **c** 22.2°
2 68.1°
3 **a** 59.1°   **b** 44.4°   **c** 68.7°
4 53.1°
5 **a** 25°   **b** 60.3°   **c** 20.9°
6 66°
7 1.2°

Practice exam questions

1 x = 416 cm or 4.16 m
2 **a** x = 13.2°   **b** w = 11 m
3 **a** x = 4.33 m   **b** y = 78.5°
4 **a** **i** *BC* = 15.6 cm **ii** *AC* = 16.7 cm
   b angle *QPR* = 43.8°
5 **a** *BC* = 2.1 m   **b** *CD* = 4.8 m
6 x = 10 cm
7 **a** x = 12.7°   **b** *BD* = 8.78 km

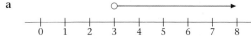

Practice questions

1 **a**
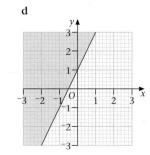

   b

   c

2 **a** −2, −1, 0, 1   **b** 2, 3, 4, 5   **c** −1, 0, 1
3 **a** $x > 3$ **b** $n \le 6$ **c** $a \le 3$
4 **a** $5.5 < x \le 8.5$ **b** $-2 \le x < 4$ **c** $-2.5 \le x < 4$
5 **a** 3, 4   **b** 1, 2   **c** −2, −1, 0, 1, 2, 3, 4, 5, 6

Practice exam questions

1 −2, −1, 0, 1, 2   2 −1, 0, 1   3 2, 3
4 $1 < x < 2$ 5 $x < -2$

Chapter 26 Graphs of linear inequalities

1 **a**   **b**
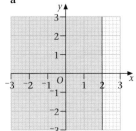

   c   **d**
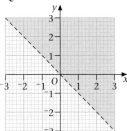

2 **a**
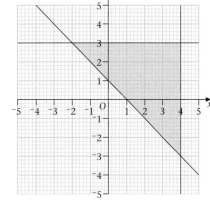

343

b

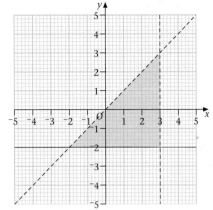

c

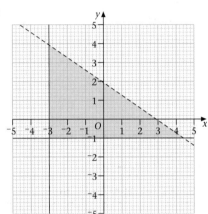

d

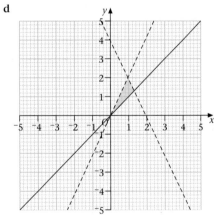

3 **a** $x \leqslant 3$ $y < 4$ $y > -2$ $y < x + 2$
   b $y \leqslant 3$ $y \geqslant x$ $y \geqslant -x$
   c $x \geqslant -1$ $y \geqslant x - 3$ $x + y \leqslant 2$
   d $x \geqslant 0$ $y \geqslant 0$ $y \geqslant 2x - 2$ $x + y \leqslant 4$ $y \leqslant x + 3$

Practice exam questions

1 **a** $x \geqslant 2$

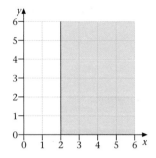

b $y \leqslant 4$

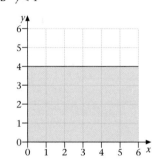

c $x + y \leqslant 4$

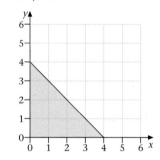

2

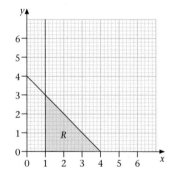

3 **a** 1, 2, 3, 4, 5
   b **i** $y = \frac{1}{2}x$
      ii $y \leqslant \frac{1}{2}x$ $x \leqslant 6$ $y \geqslant 0$

Chapter 27 Trial and improvement

Practice questions

1 $x = 3.7$ **2** $x = 2.6$ **3** $x = 2.6$ **4** $x = 3.4$

Practice exam questions

1 $x = 2.7$ **2** $x = 2.6$ **3** $x = 1.6$

Chapter 28 Constructions

Practice exam question

1

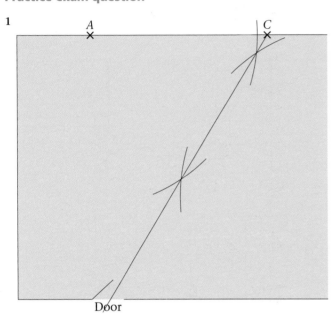

2 Construction of a right-angled triangle with hypotenuse 8 cm

Chapter 29 Loci

Practice questions

1 a

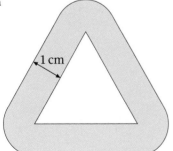

b

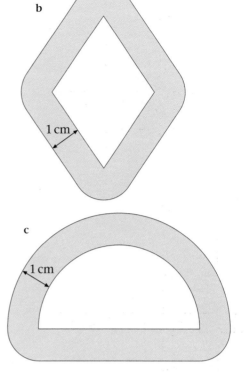

c

2

Scale 1 cm represents 1 m

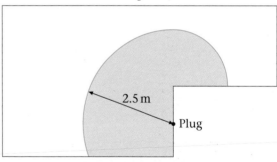

Practice exam questions

1

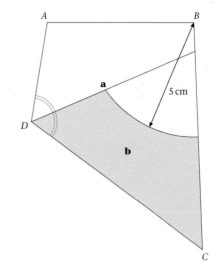

345

2

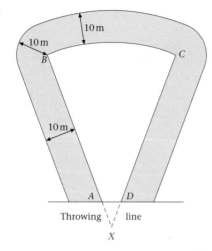

3

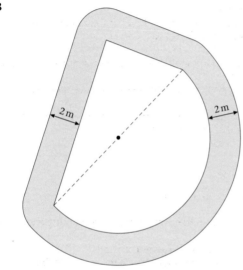

Chapter 30 Bearings and scale drawings

Practice questions 1

2 **a** 060° **b** 076° **c** 090° **d** 135°
  e 180° **f** 230° **g** 270° **h** 320°
  i 335° **j** 352°

Practice exam questions

1 **a** 5.6 miles
  b

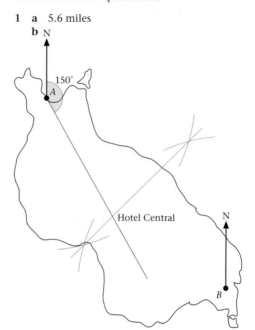

2 **a** 292°
  b

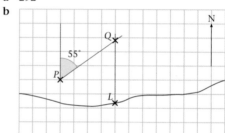

3 **a i** 037° **ii** 250°
  b 2.1 km
4 **a** 1.3 km
  b

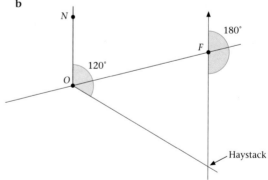

  c 25 km/h

5 a $13.9 \times 2.5 = 34.75$ km

b

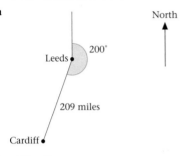

6 a

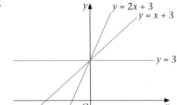

b 103 miles

7 a 333°

b $PQ = 9.22$ km, Area $PQR = 21.7$ km^2

Chapter 31 Recognising graphs

Practice question

1 a E   **b** C   **c** B   **d** A   **e** D

Practice exam questions

1 a C   **b** A   **c** E   **d** D
2 a $y = 5x$ **b** $y = 5 - x$ **c** $y = 5 - x^2$
3

[Graph showing lines labelled $y = 2x + 3$, $y = x + 3$, and $y = 3$ with axes y and x, origin O]

Chapter 32 Dimensional analysis

Practice question

1 a Length   **b** Volume
   c Length   **d** Area
   e Area   **f** Length
   g None of these   **h** Area
   i Length   **j** None of these

Practice exam questions

1 $\pi r(a^2 + b^2)$, $\frac{\pi r^2}{3}(h + 2r)$
2 a Area **b** None **c** Volume
3 a $2\pi a(a + b)$, $\frac{1}{2}(a + b)c$ **b** $\pi a^2 b$
4 a $2(v + 2w + x + y + z)$ **b** $\frac{1}{2}z(x + y)w$

Chapter 33 Similarity

Practice questions

1 a $x = 64°$
     $y = 7.5$ cm
  b $p = 6\frac{2}{3}$ cm
  c $q = 8$ cm
2 a $y = 42°$, $x = 30$ cm
  b $y = 50°$, $x = 6.43$ cm
  c $y = 50°$, $x = 60°$, $z = 39.4$ cm
3 a $y = 20$ cm, $x = 8$ cm
  b $p = 9$ cm, $q = 25$ cm
  c $a = 24$ cm, $b = 12.25$ cm
4 The rectangles are not similar because the ratio of their corresponding sides is not equal, $\frac{9}{7} \neq \frac{5}{3}$.
5 $TQ = 7.2$ cm, $PQ = 15$ cm
6 a The triangles are similar because the corresponding angles are equal.
    $\angle LMP = \angle NMO$ (vertically opposite angles are equal)
    $\angle L = \angle N$
    So the third angles in the triangles must also be equal because they sum to 180°.
  b $PM = 8.75$ cm.

Practice exam questions

1 a $x = 10.5$ cm   **b** $y = 3$ cm
2 2 cm
3 a Triangles are similar because the corresponding angles are equal.
    $\angle P$ is the same angle in both triangles
    $\angle PQR = \angle PST$ (corresponding angle on parallel lines)
    $\angle PTS = \angle PRQ$ (corresponding angle on parallel lines)
  b $QR = 7.5$ cm
4 $CD = 18$ cm

Chapter 34 Congruency

Practice questions

1 A & E   B & F   C & J   K & N   T & L
2 Triangles may be similar and not congruent i.e. one could be an enlargement of the other.
3 A & C   B & E
4 E.g. a rectangle, a kite, a parallelogram.

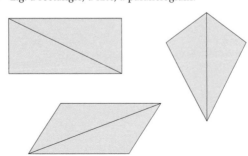

Practice exam questions

1 A & E (AAS) only these two triangles have the same corresponding side equal.
2 $x = 4$ cm, $y = 9$ cm as these are the corresponding sides.

Chapter 35 Simultaneous equations

Practice questions 1

1 $x = 3$, $y = 1$ 2 $x = 6$, $y = -1$ 3 $x = 2$, $y = 1$
4 $x = -1$, $y = 2$ 5 $x = 2$, $y = -1$

Practice questions 2

1 Adult ticket costs £3.50
2 Small bottle holds 0.5 litres; large bottle holds 2 litres
3 Gordon's score is 29

Practice exam questions

1 $x = 5$, $y = -1$ 2 $x = 2$, $y = -1$ 3 $x = 1.75$, $y = 1.25$
4 36 marks   5 $x = -3$, $y = 5$ 6 $x = 1.5$, $y = -1$

Chapter 36 Solving simultaneous equations by graph

Practice question

1 a $x = -\frac{3}{4}$, $y = -2$ b $x = 1$, $y = -1$ c $x = 3$, $y = 2$
  d $x = 3$, $y = 7$ e $x = \frac{1}{2}$, $y = \frac{1}{2}$ f $x = -1$, $y = 3$
  g $x = 3$, $y = -1$

Practice exam questions

1 a $\frac{1}{2}$
  b

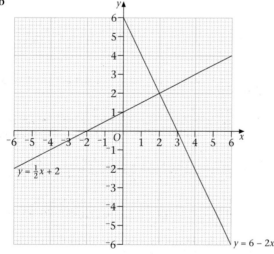

  c $x = 2$, $y = 2$

2

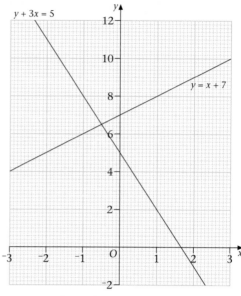

Solution: $x = -\frac{1}{2}$, $y = 6\frac{1}{2}$

Chapter 37 Volume

Practice questions 1

1 a 1080 cm^3 b 90.132 cm^3 c 403 200 cm^3
  d 34 020 mm^3 e 226 800 cm^3 f 44.88 cm^3
2 828.75 l
3 a 1000 mm^3 b 1 cm^3 c 1000 mm^3 = 1 cm^3

Practice questions 2

1 a 113 097.3 cm^3 b 850 cm^3
  c 6534.5 mm^3 d 576 cm^3
2 a 0.5969 cm^2 b 238.76 cm^3 c 2.15 kg

Practice question 3

1 a 262 cm^3 b 536 cm^3 c 39 270 cm^3 d 20.9 m^3

Practice exam questions

1 128.25 g
2 170.1 l
3 a 18 cm   b 15625 cm^3
4 No, 1 l < 1385 cm^3
5 14 000 cm^3
6 a 113.1 cm^3 b 11.03 cm^2 c 11.94 g per cm$^3.$

Chapter 38 Nets and surface areas

Practice questions 1

1

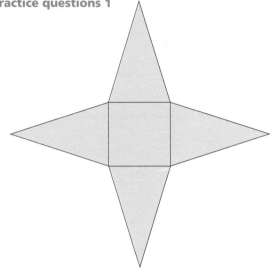

2

3 A hexagonal prism

Practice questions 2

1 a 150 cm² **b** 384 cm² **c** 1350 cm²
d 8.64 m² **e** 3456 mm²
2 a 568 cm² **b** 990 mm² **c** 5.97 m²
d 3330 cm²
3 a 1570.8 cm² **b** 9393.4 mm² **c** 75 398 cm²
4 a 2199.1 cm² **b** 12 717 mm² **c** 81 053 cm²
5 1436.7 cm²

Practice exam questions

1

Shape	Squares	Triangles	Rectangles
a Cube	6	0	0
b A squared based pyramid	1	4	0
c A triangular prism	0	2	3

2 a 4
b

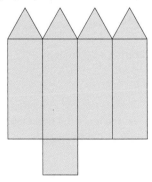

3 a Cone, prism
b Net K

4 a

b 216 cm²
5 a 150 cm²
b 125 cm³
6 a

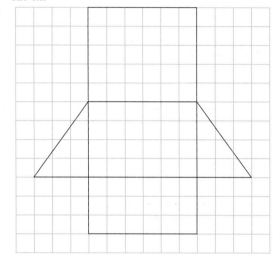

b 84 cm²

Chapter 39 Circle theorems

Practice questions

1 a 65°, angle in semicircle = 90° and angles in triangle add up to 180°
b 35°, angles on the same arc are equal
c 30°, angle at centre is twice the angle at the circumference
d 50°, angle at centre = 80°, then angles in isosceles triangle (80°, 50°, 50°)
e 15°, radius to tangent = 90°, then isosceles triangle
f 112°, opposite angles of a cyclic quadrilateral add up to 180°
g 56°, isosceles triangle and radius to tangent = 90°
h 58°, congruent triangles, radius to tangent = 90° and angles in triangle add up to 180°
2 a 33° **b** 34° **c** 74° **d** 68°
e 63° **f** 112° **g** 74° **h** 38°

Practice exam questions

1 a 56°, angle at centre is twice the angle at the circumference
b 124°, opposite angles of a cyclic quadrilateral add up to 180°
2 a 24°, angles on the same arc are equal
b 86°, opposite angles of a cyclic quadrilateral add up to 180°

3 Angle $CDA = 110°$, opposite angles of a cyclic quadrilateral add up to 180°
Angle $EDA = 70°$, angles on a straight line add up to 180°
Angle $DAE = 60°$, angles in a triangle add up to 180°
4 a Angle $ABC = 58°$
b Angle $OCB = 26°$, radius to tangent $= 90°$
Angle $OCA = (180 - 116) ÷ 2 = 32°$
Angle $ACB = 32° + 26° = 58°$
Angle $ACB =$ angle ABC, therefore triangle ABC is isosceles

Chapter 40 Proof

Practice questions

1 a $(x + 2)^2 = (x + 2)(x + 2)$
$= x^2 + 2x + 2x + 4$
$= x^2 + 4x + 4$
b $(x^2 + 4x + 4)/(x^2 + 2x) = (x + 2)(x + 2)/x(x + 2)$
$= (x + 2)/x$
2 a

b

c

3 Base angles of the isosceles triangle are equal.
Let each base angle $= y$
$2x + y + y = 180$ (angle sum of a triangle)
$2x + 2y = 180$
$2y = 180 - 2x$
$y = 90 - x$
So the interior angle at B is $90 - x$
Exterior angle $= 180 -$ interior angle
So exterior angle at $B = 180 - (90 - x)$
$= 180 - 90 + x$
$= 90 + x$
4 Using a triangle ABC as shown.

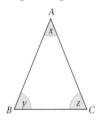

Draw a line through A parallel to BC.
We now have two alternate angles at A as shown.

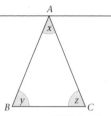

So $x + y + z = 180°$ (Angles on a straight line)

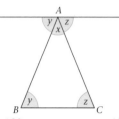

$z = 180 - x - y$ \qquad (Angle sum of a triangle)

5

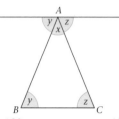

$w = 180 - (180 - x - y)$ \quad (Angles on a straight line)
$w = 180 - 180 + x + y$
$w = x + y$
So exterior angle $w =$ sum of interior opposite angles $(x + y)$

Practice exam questions

1 a $10 \times 5 = 50$
b 30 cm
2 Yes, she is correct.
Area of 1 square $= 5 \times 5 = 25$ cm^2
Area of 2 squares $= 50$ cm^2
Area of triangle is less than area of square
Therefore, total area of shape is less than $25 + 25 + 25$ cm^2
(75 cm^2)

Practice exam paper

Paper 1

1 a 32 \qquad **b** 10
2 a 15 \qquad **b** $9a - 2b$
3 16.00
4 21 cm
5 a B marked on correct bearing 4 cm from A.
b 270° \qquad **c** 4.8 cm
6 a 40° \qquad **b** 108°
7 32 cm^2
8 a i 5, 8, 11
ii Valid explanation e.g. Sequence is 3 times table + 2.
Or 29 and 32 are in the sequence.
b $3n + 1$
9 15 years old

10 a

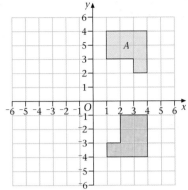

b i Reflection in the line $x = -1$

ii $\begin{pmatrix} -1 \\ -5 \end{pmatrix}$

11

12 a 10   **b** 36°

14 a

x	1	2	3	4	6	12
y	12	6	4	3	2	1

b

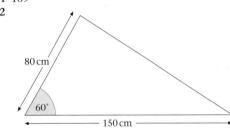

13 a $13x + 14$ **b** $x^3 + 4x$
   c 1.7

15

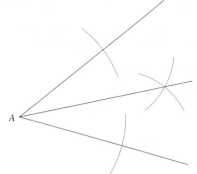

16 450 mm^2
17 a −3   **b** 4
18 $x = y^2 - 4$
19 a 28°   **b i** 148°   **ii** 106°
20 $x = 3, y = 1$

Paper 2

1 189
2

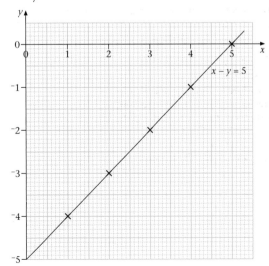

3 a $x = 4$
  b $y = 7$
4 a 23.04 cm^2
  b The rectangle is twice as long as the square. Since it has the same area then the height must be half as much as the height of the square.
     Height of rectangle = 4.8 ÷ 2 = 2.4 cm
5 30.73 cm^2
6 a (1, −4), (2, −3), (4, −1), (5, 0)
  b $x - y = 5$

  c Q is above the line.

7 $x = 88°$

8

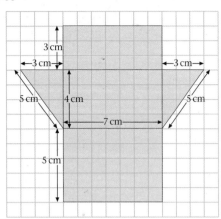

9 a

n	$n + 1$
$n + 10$	$n + 11$

  b $4n + 22$

  c n is an integer. $4n$ will always be an even integer as it will always divide by 2.
$4n + 22 = 2(2n + 11)$ is also an even integer as it will always divide by 2.

10 8π or 25.13 cm (2 dp).

11 $x = 28°$

12

x	$x^3 - 2x$	Comment
2	4	too small
3	21	too big
2.5	10.625	too big
2.4	9.024	too small
2.45	9.806125	too small

Then x must be greater than 2.45 but smaller than 2.5.
So $x = 2.5$ to 1 decimal place.

13 If Wayne is correct the triangle will be right-angled and Pythagoras' theorem will work.
$(9.1)^2 + (12.5)^2 = 239.06$
$\sqrt{239.06} = 15.46156525$
This does not equal 15.3 and so Wayne is not correct.
The triangle is not right angled.

14 11.78 m^2

15 $x = 7$

16 a length
  b volume
  c none of these.

17 a $-2, -1, 0, 1, 2, 3, 4$
  b $y = x + 1$
  c

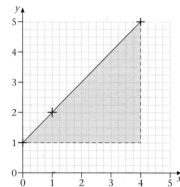

18 4.59 m (2 d.p.)

19 a $(x + 4)(x - 9)$
  b $x = -4$ or $x = +9$